Kraftstoffe für die Mobilität von morgen

5. Tagung der Fuels Joint Research Group
am 30. Juni und 01. Juli 2022 in Waischenfeld

KRAFTSTOFFE FÜR DIE MOBILITÄT VON MORGEN

Herausgeber: Jürgen Krahl, Axel Munack, Peter Eilts, Jürgen Bünger

Cuvillier Verlag Göttingen

Bibliografische Information der Deutschen Nationalbibliothek
Die Deutsche Nationalbibliothek verzeichnet diese Publikation in der Deutschen Nationalbibliografie; detaillierte bibliographische Daten sind im Internet über http://dnb.d-nb.de abrufbar.
1. Aufl. - Göttingen: Cuvillier, 2022

Buchcover Bildnachweis: photocase.com / Markus Gann, photocase.com / crocodile

Nonnenstieg 8, 37075 Göttingen
Telefon: 0551-54724-0
Telefax: 0551-54724-21
www.cuvillier.de

1. Auflage, 2022
Gedruckt auf umweltfreundlichem, säurefreiem Papier aus nachhaltiger Forstwirtschaft.

ISBN 978-3-7369-7624-5
eISBN 978-3-7369-6624-6

Inhaltsverzeichnis

FVV Fuels Study IV – Transformation of European Mobility to the GHG Neutral Post Fossil Age

Ulrich Kramer, David Bothe, Frank Dünnebeil

Abstract

On the basis of the FVV Fuels Study IV [1], different technology (energy/fuel – powertrain) pathways (100% scenarios) are investigated that all lead to carbon neutral mobility in 2050. They are compared with regard to energy demand, total costs, cumulative greenhouse gas emissions and other environmental impacts by means of a holistic cradle-to-grave approach, that considers all relevant costs and emissions from vehicle production and the setup of a complete sustainable energy supply system. For this purpose, the emissions caused by the construction of the infrastructure, such as wind turbines, electrolysers or charging stations, are also included. Only such an analysis of the entire energy system delivers valid insights into efficient defossilisation strategies.

Regardless of which of the technology pathways Europe chooses, the 1.5 °C greenhouse gas budget discussed for Europe will already be exceeded by 2032 due to transport emissions alone, when the ramp-up of carbon neutral technology is determined by the vehicle fleet exchange rate of new carbon neutral vehicle technology. The reason for this is the dominant share of the existing fleet still using fossil fuels, phasing out by 2050. Irrespective of the scenario, the GHG emissions of the existing fleet amount to around 70 % at an identical rate of introduction of new vehicle technologies and accordingly replacing fossil energy in transportation at the same rate. Without technology options that reduce emissions in the existing fleet and an efficient combination of defossilisation measures, it will not be possible to achieve the ambitious European climate targets.

Therefore, a follow-up study has been started to thoroughly investigate the constraining ramp-up bottlenecks ("FVV Fuels Study IVb").

First potential technical bottlenecks could be identified which are likely to decelerate the ramp-up of renewable mobility, sticking to exclusive 100%-scenarios, even under absolute ideal legal boundary conditions and investment attractiveness. These bottlenecks may involve the achievable ramp-up speed of key technologies as electrolysis, direct air capturing, reverse water gas shift reaction, charging infrastructure, electrical grid extension or build capacity of solar and wind power generation. To what extent these bottlenecks are limiting the required fast ramp-up of complete sustainable transportation pathways has not been finally investigated yet. Final results are expected by autumn 2022.

However, interim results turn more and more out to confirm, that concentration on a single energy-technology pathway is limiting the achievable ramp-up speed significantly. Most efficient GHG reduction requires intelligent technology mixes that enable the fastest possible exit out of fossil energy use for the lowest costs.

Kurzfassung

Auf der Grundlage der FVV-Kraftstoffstudie IV werden verschiedene Technologiepfade (Energie/Kraftstoff - Antriebsstrang) (100%-Szenarien) untersucht, die alle zu einer klimaneutralen Mobilität im Jahr 2050 führen. Sie werden hinsichtlich des Energiebedarfs, der Gesamtkosten, der kumulierten Treibhausgasemissionen und anderer Umweltauswirkungen mit Hilfe eines ganzheitlichen Cradle-to-Grave-Ansatzes verglichen, der alle relevanten Kosten und Emissionen aus der Fahrzeugproduktion und dem Aufbau eines vollständigen nachhaltigen Energieversorgungssystems berücksichtigt. Dabei werden auch die Emissionen einbezogen, die durch den Aufbau der Infrastruktur, wie Windkraftanlagen, Elektrolyseure oder Ladestationen, entstehen. Nur eine solche Analyse des gesamten Energiesystems liefert valide Erkenntnisse über effiziente Defossilisierungsstrategien.

Unabhängig davon, für welchen der Technologiepfade sich Europa entscheidet, wird das für Europa diskutierte 1,5 °C-Treibhausgasbudget bereits im Jahr 2032 allein durch die Emissionen des Verkehrs überschritten. Der Grund dafür ist der dominierende Anteil der bestehenden Flotte, die noch fossile Kraftstoffe verwendet und bis 2050 ausläuft. Unabhängig vom Szenario belaufen sich die THG-Emissionen der bestehenden Flotte auf etwa 70 %, wenn die Einführung neuer Fahrzeugtechnologien gleich schnell erfolgt und dementsprechend fossile Energie im Verkehr in gleichem Maße ersetzt wird. Ohne Technologieoptionen, die die Emissionen der bestehenden Flotte reduzieren, und eine effiziente Kombination von Defossilisierungsmaßnahmen werden die ehrgeizigen europäischen Klimaziele nicht erreicht werden können.

In einer schon begonnenen Folgestudie werden daher Engpässe von Hochläufen zurzeit gründlich untersucht.

Erste potenzielle technische Engpässe konnten identifiziert werden, die den Hochlauf erneuerbarer Mobilität bei einem Festhalten an ausschließlichen 100%-Szenarien auch unter absolut idealen rechtlichen Rahmenbedingungen und gesicherter Investitionsattraktivität weiter verlangsamen dürften. Diese Engpässe können die erreichbare Hochlaufgeschwindigkeit von Schlüsseltechnologien wie Elektrolyse, CO_2-Abscheidung aus der Umgebungsluft, inverse Wasser-Gas-Shift-Reaktion, Ladeinfrastruktur, Stromnetzerweiterung oder Kapazitätsaufbau bei der Solar- und Windenergieerzeugung betreffen. Inwieweit diese Engpässe den notwendigen schnellen Hochlauf vollständig nachhaltiger Transportwege einschränken, ist noch nicht abschließend untersucht. Endgültige Ergebnisse werden bis Herbst 2022 erwartet.

Zwischenergebnisse bestätigen jedoch mehr und mehr, dass die Konzentration auf einen einzigen energietechnischen Pfad die erreichbare Hochlaufgeschwindigkeit erheblich begrenzt. Eine möglichst effiziente THG-Reduktion erfordert intelligente Technologie-Mixe, die den schnellstmöglichen Ausstieg aus der Nutzung fossiler Energien zu den niedrigsten Kosten ermöglichen.

1. Aim and Method

The EU plans to reach full climate neutrality across all sectors by 2050. For the transport sector in Europe, this aim cannot be achieved with combustion engine powered vehicles using fossil fuels. To reach a carbon neutral transport sector and meet both, national and European CO_2 targets, appropriate concepts for the defossilisation of the transport sector are required.

To investigate how this goal can be reached, the FVV working group "Fuels" has compared and evaluated different mobility scenarios which will allow fully carbon neutral mobility in 2050 (including the whole fuel supply chain as well as vehicle production) and for which energy demand will solely be supplied by renewable wind and solar energy.

This study illustrates various "energy and drivetrain technology pathways", all of which have the potential to defossilise the transport sector by 2050. All of the fuel / drivetrain pathways are evaluated in so called "100% scenarios", where every segment of the transport sector is assumed to be powered by the respective technology if technically feasible. Interaction of the transport with other sectors is not part of the study. These extreme scenarios are theoretical and not meant to be a realistic forecast of future developments. However, they allow for a comprehensive comparison across different fuel / drivetrain pathways and illustrate potential challenges arising from industry level scale up. The considered fuel / drivetrain pathways are not based on any fossil sources. Local CO_2 emissions are allowed if they are fully compensated during the production process (e.g., capturing CO_2 directly from the air, closed CO_2 circle).

The focus of the study is a quantitative and qualitative comparison of mobility costs (including the costs for the energy/fuel production and distribution facilities as well as vehicle costs), primary energy demand (including losses along the complete energy/fuel supply chain), environmental impacts (especially greenhouse gases) and critical raw materials (e.g., lithium). Thereby, all relevant phases of the lifecycle are taken into account, including the production of vehicles as well as the required incremental build-up of the necessary energy/fuel supply infrastructure (energy provision and distribution).

The modelled energy provision is CO_2 neutral and solely provided by wind and solar energy. The renewable energy is then used in seven different energy pathways:

- 1 pathway: Direct use in battery electric vehicles and catenary grid supplied long haul trucks ("BEV")
- 2 pathways: Producing hydrogen via electrolysis which then is used in vehicles that either are equipped with a fuel cell (fuel cell electric vehicles, "FCEV") or with an internal combustion engine optimized to combust hydrogen ("H_2-Comb")

- 4 pathways: Producing so-called Power-to-X (PtX) fuels by again producing hydrogen via electrolysis, capturing CO_2 directly from the air (DAC) and then finally the synthesis of Methane, Methanol ("MeOH"), Dimethylether ("DME") or Fischer-Tropsch-fuels ("FT-gasoline/diesel")

The starting point of the analysis is the total mobility demand and its development until 2050, based on an EU Reference scenario. We then proceed to derive the required future development of the vehicle fleet (for the road sector) and new registrations for different vehicle segments to achieve 100% fleet penetration with the respective defossilised drivetrain concept in 2050. The so modelled fleet development enables combinations with annual mileages and specific energy efficiencies and thus allows us to determine the energy/fuel demand for the road sector in all scenario years. For inner European Rail, Aviation and Shipping, we use a simplified approach, as their relevance is subordinate.

In all of our scenarios, we assume that all new vehicles with alternative powertrains are fully operated with additionally generated renewable energy. This is also assumed for identical shares of new vehicles operated with Fischer-Tropsch gasoline/diesel for comparability reasons, even if this fuel is compatible to the whole existing vehicle fleet. The total renewable energy/fuel demand (or Tank to Wheel (TtW) demand) is then the starting point for our energy/fuel supply chain modelling. Following a bottom-up approach, we trace the energy demand across the different steps of each supply chain to determine the required build-up of capacities of each element (such as electrolysis or power generation) over the course of time. We focus our analysis solely on the renewable energy supply for transport, without any interactions with other sectors. A global transformation of energy supply and industrial processes to fossil-free alternatives is also assumed for the technologies used to build the infrastructure. The resulting greenhouse gas emissions of these processes are assumed to be reduced to a minimum in 2050 (transformation of the "background system" of material supply and production processes to become "quasi GHG free" from 2021 to 2050).

All required steps of the supply chain including generation, transport and storage are considered (see Figure 2 for an exemplary illustration of a modelled supply chain). Once the infrastructure and energy/fuel requirements have been assessed, they are evaluated across the different dimensions outlined above – environmental impacts, material demand and costs.

Several aspects of future development, particularly with respect to future vehicle technologies and the future sourcing of the required energy, are currently uncertain and subject to various factors, particularly technical, political and regulatory decisions. To reflect this uncertainty, we assess three levels of future technological development of vehicles (labelled "Status Quo", "Balanced" and "All-In") and two places of energy sourcing (Europe and Worldwide) in our analysis of the seven different energy / drivetrain pathways. This results in a total of 42 different scenarios assessed.

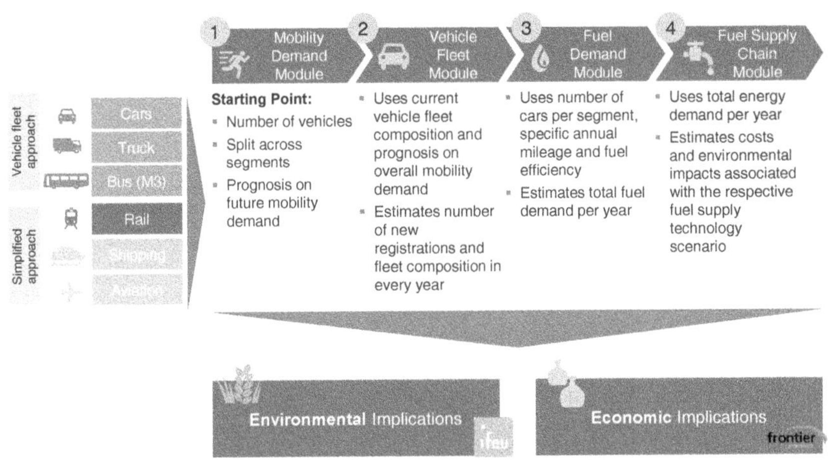

Fig. 1: Schematic overview of our modelling approach [Source: Frontier Economics, ifeu]

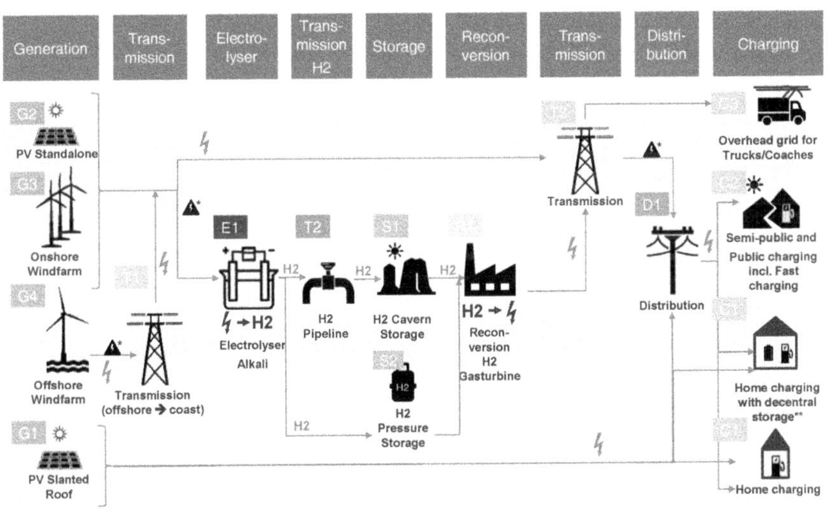

Fig. 2: Overview of level of detail of supply chain model (here shown for BEV) [Source: Frontier Economics, ifeu]

Compared to the previous fuel study of FVV (FVV Fuels Study III, 2018) [2], which modelled 100% scenarios for various energy / drivetrain pathways for the road mobility in Germany for the "photo year 2050", this study expands the geographic scope to

EU27+UK and focuses on all transport sectors, while at the same time including a more detailed breakdown for the road sector. All scenarios are simulated for the "photo years" 2020, 2030 and 2050, in order to describe the ramp-up from today into a defossilised future 2050. The analysis is further complemented by an economic and environmental assessment, covering all phases of a vehicle life and the provision of final energy carriers, including the required infrastructure (e.g., for energy/fuel generation, transport, storage, and distribution). The focus is solely on the transport sector – potential interactions with other sectors (i.e. sector coupling) are not taken into account.

2. Energy Demand and required capacities in 2050

The required total energy in the mobility sector (on a Well-to-Wheel basis, taking into account the losses along the energy/fuel supply chain) determines the requirements for initial generation capacities (PV and wind plants), as well as any infrastructure requirements further down the supply chains. Relative comparisons of the WtW energy demand across the different energy / drivetrain pathways are therefore a valuable indication for further assessments. Figure 3 summarizes the results for the 42 different fuel / drivetrain pathways: BEV by far requires the lowest WtW energy demand (starting from 2,000 TWh, which is around 68% of EU27+UK electricity demand in 2019), due to its low TtW demand. The highest WtW demand is required for synthetic fuels (up to 10,000 TWh), due to higher TtW demands and high losses along the energy/fuel supply chain. These are particularly driven by the fuel synthesis and electrolysis.
Hydrogen powered fuel cell vehicles (H_2-FCEV) require approximately twice as much WtW energy as BEV, while H_2 powered vehicles equipped with combustion engines (H_2 Comb) consume approximately 2.5 to 3 times as much energy as BEV. Sufficient amounts of legacy fleet compatible, defossilised Fischer-Tropsch diesel/gasoline require 3.5 to 4 times as much WtW energy as BEV.

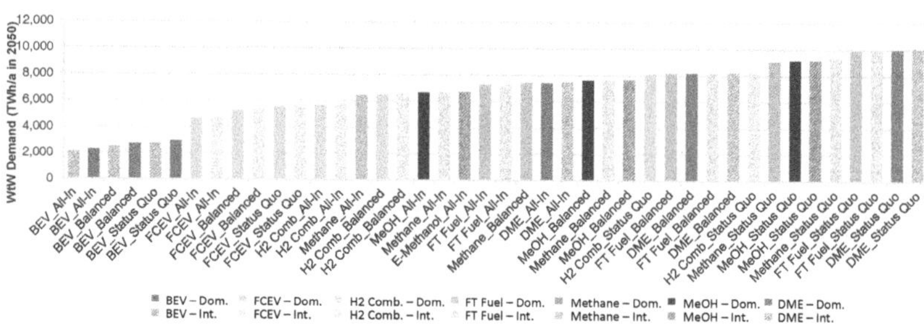

Fig. 3: WtW Demand in TWh/a in 2050 for 42 scenarios [Source: Frontier Economics]

However, for environmental impacts as well as for costs of the energy/fuel supply chain, not the WtW energy demand, but the required installed capacities are the deciding factor. Figure 4 therefore summarizes the required capacities of renewable energy generation infrastructure for all scenarios (for the road sector). Without exception, the domestic energy sourcing scenarios require much higher generation capacities than international scenarios, where energy is also sourced from regions outside of Europe such as MENA or Patagonia. This is due to the fact that regions outside of Europe have better conditions for generating renewable energy (e.g., hours of sunshine and/or wind). The highest generation capacities are required for domestically produced synthetic fuels, as FT diesel/gasoline or DME (up to 4,800 GW), while BEV scenarios require the lowest generation capacities (starting from 750 GW when energy is sourced internationally from MENA, and from 1,100 GW for domestic energy sourcing). By way of comparison, 340 GW of renewable wind and solar generation are currently installed in Europe for all sectors, which is planned to be increased to up to 690 GW by 2030. The factor of required installed power generation capacity for "FT-ICE / BEV" is in the range of 3 for domestic energy sourcing. When FT fuel is produced internationally the factor is reduced to approximately 2 to 2.5.

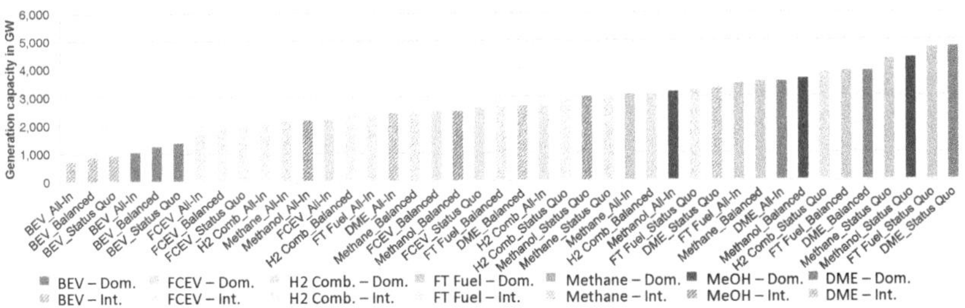

Fig. 4: Generation capacity in GW in 2050 for all 42 scenarios
[Source: Frontier Economics]

Similarly, hydrogen plays a role in all scenarios, albeit in varying forms. Electrolysis is thus a key element for a carbon neutral mobility sector, independent from the selected energy pathway. All fuel / drivetrain pathways require significant electrolysis capacities, including BEV, as in a fully renewable energy system a chemical buffer (here: hydrogen) is required for dark doldrums to buffer seasonal fluctuations. In the domestic scenario for the road sector, the required installed electrolysis capacity ranges from 870 GW up to 2,200 GW in 2050, solely for transportation. Currently, only 40 GW are planned for EU27+UK until 2030. H_2-FCEV pathways finally (in 2050) require approximately 1,200 GW, H_2 Comb 1,600 GW and FT-ICE 1,900 GW of electrolysis capacity. In the BEV scenarios, approximately 600 GW (international) and 1,000 GW (domestic) of electrolysis capacities need to be built until 2050, in order to maximize the utilization of all renewable power generated. The ramp-up of electrolysis capacity is therefore likely to become a temporary bottleneck.

3. Environmental impacts

With a full defossilisation of the transport sector by 2050, annual GHG emissions are in all energy / drivetrain pathways 95-97% lower than in the baseline year 2020. Origin of the small amount of remaining unavoidable GHG emissions are primarily processes in the background system (as e.g., concrete use for wind turbine foundations, methane slip). However, the contribution of the transport sector to global warming depends on its cumulative emissions over the entire pathway towards full defossilisation. Assessing the GHG mitigation effectiveness of different defossilisation pathways must therefore include the GHG emission backpack associated with the ramp-up to a 100% defossilised transport sector. In our methodology with 100% back-casting scenarios, cumulative GHG emissions 2021 to 2050 turn out to be in a comparable order of magnitude in all scenarios (bandwidth of road transport in the range of 14%). This is mainly due to the assumed identical ramp-up speed of alternative vehicle concepts (determined by the assumed vehicle fleet exchange rate) and renewable energy/fuel supply required for achieving 100% of the respective pathway in the year 2050. Cumulative emissions in all pathways are dominated by operation of the remaining gasoline/diesel vehicle fleet with fossil fuels (already containing 7% biofuel share) with a total contribution of 66-74%, as 100% defossilised energy/fuel supply will be achieved only in 2050. The ramp-up of renewable energy/fuel supply chain infrastructure contributes 5-20% and vehicle production 11-24% to total cumulative GHG emissions.

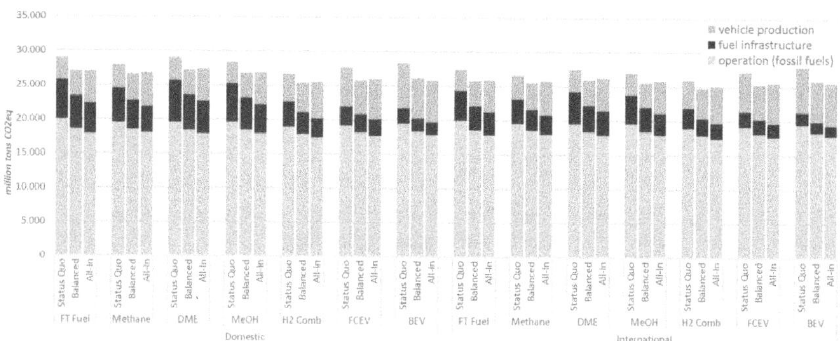

Fig. 5: Cumulative GHG emissions in all 100% scenarios with identical ramp-up speed of defossilisation [Source: ifeu]

In all 100% scenarios, we assume a linear ramp-up of new registrations of alternative drivetrain technologies up to 100% sales share at a point of time, which allows a complete fleet renewal until 2050 ("backcasting" approach). For passenger cars and light duty vehicles 100% sales share of the defossilised drivetrain technology is required in 2033, for heavy duty trucks it is later due to shorter lifetimes (e.g., in 2042 for long haul). Complete fleet turnover with new vehicle technology and associated build-up of defossilised energy/fuel supply chain infrastructure until 100% is achieved in 2050. The

same ramp-up speed is also assumed for legacy fleet compatible FT gasoline/diesel, even if this fuel could already be used in existing gasoline/diesel vehicles.
In reality, however, actually reachable ramp-up speeds will most probably differ considerably between the technology pathways. A sensitivity analysis shows that ramp-up speed of defossilised final energy supply is the crucial factor determining how fast GHG emissions of the transport can be reduced with purely technical measures (without transport-reducing and modal-shift measures) and which cumulative GHG emissions are to be expected over the entire transition period. As investigated on the example of the FT fuel pathway, the achievable ramp-up speed has a significant greater impact on cumulative GHG emissions than the choice of the pathway itself, if identical ramp-up speeds for all pathways are assumed. Quickest possible applicability of substantial quantities of renewable energy to reduce dependencies on fossil fuels is essential for minimizing GHG emissions from transport and, therefore, measures applied already in the present decade are most important for the reduction of the GHG backpack until 2050.

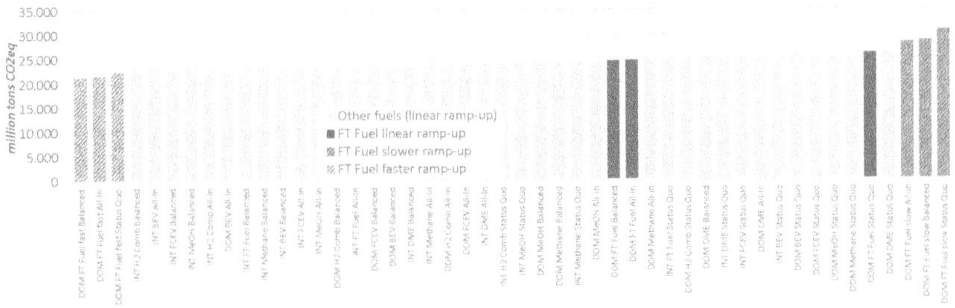

Fig. 6: Sensitivity analysis for the impact of different market ramp-up speeds for FT fuels in road transport on cumulative GHG emissions 2021-2050 associated with the EU27+UK road transport [Source: ifeu]

In all energy / drivetrain pathways, better vehicle efficiencies in the “Balanced” technology scenarios lead to lower cumulative GHG emissions compared to the “Status Quo” technology scenarios. However, the “All-in” scenarios with highest vehicle efficiencies lead to slightly increased cumulative GHG emissions compared to the “Balanced” scenarios for the FCEV and all ICE pathways. The additional GHG from vehicle production with aluminum light weighting outweighs GHG savings from efficiency improvements. Thus, segregated energy efficiency optimization per sector is not necessarily leading to the most efficient solution for overall GHG reduction.
The international energy/fuel supply scenarios deliver the lowest cumulative GHG emissions. The savings by international energy/fuel sourcing compared to local sourcing (only in Europe) are 1% - 2% for BEV, 2% - 3% for H_2 pathways (FCEV, H_2 Comb) and 4% - 6% for hydrocarbon e-fuel pathways.
In order to assess compatibility of the 100% scenarios with the Paris climate targets, we compare cumulative GHG emissions with estimates of the remaining CO_2 budget

for the European Union. In all 100% scenarios, the GHG emissions associated with the transport sector (including vehicle production and defossilised energy supply infrastructure) will exceed the total 1.5 °C GHG budget for Europe (EU27+UK, all sectors, 67% probability) in 2031 - 2032 and will require 43% - 51% of the total 1.75 °C GHG budget (50% probability) for Europe. This indicates that an exclusively technical defossilisation with one single energy / drivetrain pathway and assumed vehicle characteristics cannot meet the GHG reduction requirements on Europe's transport sector. Further GHG reduction potentials need to be analyzed and applied urgently.

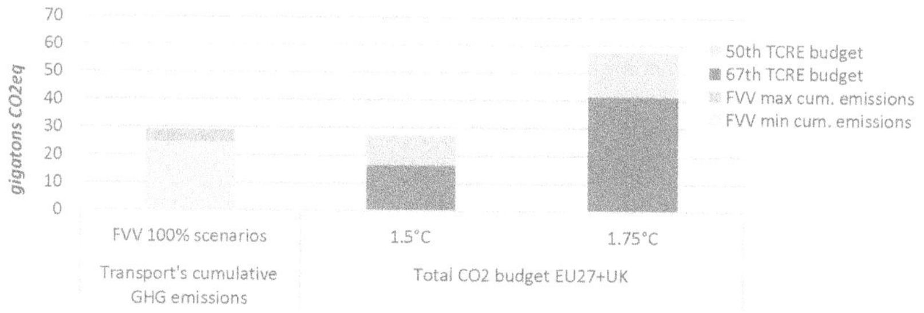

Fig. 7: Comparison of cumulative GHG emissions of EU27+UK transport 2021-2050 with total CO_2 budgets (own estimates for all sectors) of EU27+UK 2021-2050 [Source: ifeu]

Further environmental impact categories considered in this study (acidification, eutrophication, PM formation) do not show general ecological risks for any of the defossilisation pathways. Eutrophication and PM formation potentials show a strong reduction from 2020 to 2050 for all pathways. Annual acidification potential is reduced from 2020 to 2050 by 30-50 % in the H_2-FCEV and all ICE scenarios. Since contribution of land-based transport to acidification is very low, even a slight increase of acidification potential in the BEV Status quo scenario would not cause an environmental bottleneck. Furthermore, land use for renewable power generation for the defossilised transport sector does not generally pose an ecological bottleneck. For the domestic energy sourcing scenario, we assumed that Europe could become energy independent. As laid out in other studies this also depends on the development of key technologies such as "floating offshore wind". In the domestic energy sourcing scenario land use for power generation for European transportation requires 0.5% to 1.3% of EU27+UK land area, which corresponds to an area up to twice the size of Belgium. International energy sourcing requires about 1/3 less land use than energy sourcing exclusively in Europe. Land use of all other facilities in the defossilised energy/fuel supply chain (DAC, synthesis plants etc.) is negligible (e.g., DAC land use is max. 0.004% of EU27+UK land area). However, installation of renewable power generation capacities should avoid environmentally sensitive areas in order to minimize land use related environmental impacts.

4. Rare materials

In all pathways for the defossilisation of the transport sector, availability of selected raw materials can be a limiting factor for a fast market ramp-up and for achieving 100% in 2050, if a fair share of Europe on the raw material demand for a worldwide defossilisation is assumed. Identified bottlenecks result mainly by specific vehicle configurations in the 100% scenarios and by the assumed strong future motorization increase in non-European countries in the theoretical case of a worldwide economic catch-up to the European Union by 2050.

A worldwide ramp-up of electric mobility can be affected by absolute and temporary lithium or cobalt bottlenecks. With the specific battery configurations assumed in our scenarios (Li-NMC battery technology, as state of the art in Europe), extrapolated worldwide material demand is on the upper end impeding 100% worldwide electric mobility. A future mix of different battery technologies, with lower lithium and cobalt content could reduce raw material demand. Furthermore, global lithium and cobalt resources and reserves have developed very dynamically in the last few years. Therefore, a considerable future increase of primary material supply can be expected.

Platinum is a clear bottleneck in the FCEV scenarios. Global platinum supply could only fulfil the demand of Europe's transport sector for 100% FCEV. Assuming similar developments of FCEV fleets in the rest of the world, global demand will clearly exceed currently known reserves and lead to absolute bottlenecks. In all other fuel / drivetrain pathways currently known reserves of platinum group metals (PGM) are sufficient to fulfil cumulative demand for primary PGM for the global mobility sector in all scenarios. Further materials such as copper, silver, nickel, and neodymium are required in vehicle production and / or the energy/fuel supply chain infrastructure and could therefore cause bottlenecks in all energy/fuel pathways. However, proactive demand and supply strategies can prevent future bottlenecks of these materials: Primary material demand can be reduced in transport and other demand sectors by increase of recycling, substitution with less critical materials or use of existing alternative technologies. At the same time, supply has to be increased based on sustainable mining and supply systems.

5. Costs

While costs do not constitute a binding constraint, it is of common interest for consumers, manufacturers and governments to proceed with an economical pathway to transform the transport sector. Identifying core cost drivers and dependencies can also aid in determining which technological measures might have a particularly good cost benefit ratio.

We have looked into vehicle and energy/fuel supply chain costs. The results are expressed in terms of incremental costs – so additional investments that need to be made compared to today. This approach focuses on the differences between various fuel / drivetrain pathways and thereby provides a clearer picture of the cost effects of choosing different drivetrain technologies. Total incremental costs are expressed in 2020 Euros and as Net Present Values. The total incremental costs to defossilise the complete European (EU27+UK) road transport sector by 2050 range from 2,600 billion € to 5,300 billion €, which corresponds to 17% to 34% of the annual European (EU27+UK) GDP in 2020 (15,600 billion € in 2020). Figure 8 compares the total incremental costs for all 42 scenarios.

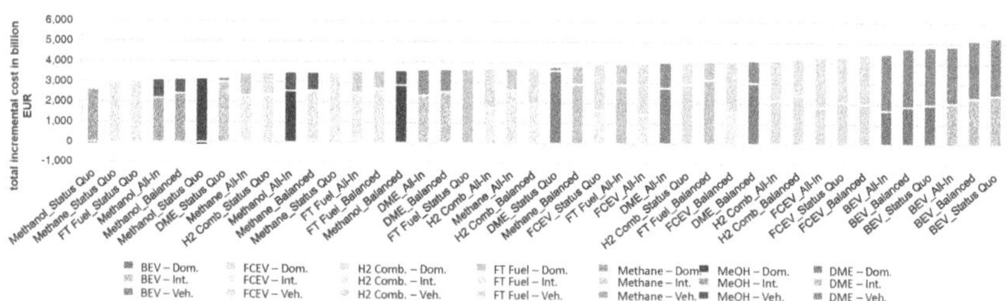

Fig. 8: Total incremental costs – energy/fuel supply chain and vehicles – until 2050 in billion € [Source: Frontier Economics]

The highest total incremental costs (in NPV) for total road transport are found in the BEV scenarios (4,500-5,300 billion €) followed by FCEV (3,900-4,500 billion €), where costs for passenger vehicles dominate overall costs. BEV vehicle costs are driven by the battery costs, determined by range assumptions (as e.g., 300 – 500 km passenger car/LDV range) and resulting battery sizes, as well as assumed specific battery costs (160 €/kWh for 2020, 120 €/kWh for 2030, 80 €/kWh for 2050). The BEV vehicle costs are modelled to decrease over time with assumed progress in battery technology development. The incremental vehicle costs dominate the total costs for the BEV and FCEV pathways and contribute to 50% of the total costs. The lowest total incremental costs for road transport are found for synthetic fuels, particularly due to lower incremental vehicle costs.

Energy generation costs are the main driver of energy/fuel supply chain costs. The total energy/fuel supply chain costs are lower in the international than in the domestic scenario for most fuels, driven by a decrease in total generation costs. Not only are the specific power generation costs per unit assumed to be lower outside than within Europe, but at the same time, fewer generation capacities are required overall due to better conditions for wind and/or solar generation (higher full-load hours). Only the international BEV scenario is expected to be more expensive than the domestic scenario because of expensive import power lines to Europe. When assessing pathways using international energy sourcing, significant political uncertainties have to be considered affecting the general feasibility as well as cost levels.

The lowest total incremental costs for total road transport are found for "E-Fuel-ICE international sourcing pathways" with continued 2020 vehicle technology ("Status-Quo" pathway: without hybridization or light-weight measures), starting from Methanol-ICE (2,600 billion €) over FT-gasoline/diesel-ICE (3,000 billion €) up to H2-Comb (3,500 billion €).

The cumulated incremental costs for the road sector of the "Status Quo vehicle technology scenarios" are lower than those for the "Balanced scenarios" and "All-In scenarios" for all fuel/drivetrain pathways, except for BEV. Vehicle cost increases from hybridization or light-weight measures outweigh cost savings from reduced energy/fuel supply infrastructure requirements. When considering the cumulative total costs, it is more cost efficient to build additional power generation and energy/fuel distribution infrastructure than to maximize efficiency measures (at high cost) on the vehicle level. Neither hybridization, nor light weight measures are paying off when cumulative total costs are compared. Therefore, stringent sector limits, such as TtW GHG targets for vehicles, can lead to increased cumulative total costs. Energy efficiency optimization per sector does not necessarily lead to the most cost-efficient solution for GHG reduction.

6. Main Conclusions

The defossilisation of the complete European (EU27 + UK) transport sector is possible and affordable. The total costs are in a range of 1% of the European GDP per year over 30 years, with incremental vehicle costs dominating the total defossilisation costs for BEV and FCEV pathways and costs for energy/fuel supply chain infrastructure dominating the total defossilisation costs for synthetic fuels. Therefore, vehicle costs as well as energy/fuel supply chain infrastructure costs must be considered in any economic system optimization and GHG reduction strategy. The lowest total incremental costs for passenger cars (incl. LDV) are achieved with e-fuel operated ICEV with continued 2020 vehicle technology (without hybridization or light-weight measures). Total costs are lower for international energy sourcing than for domestic energy sourcing for all ICEV and FCEV pathways, while domestic costs are lower for the BEV scenario. Therefore, mixed scenarios with domestic BEV and international FCEV/ICEV energy sourcing appear to be the most cost efficient.

The introduction of alternative vehicle technology pathways is limited by the vehicle fleet exchange rate. In all investigated 100% scenarios - with assumed defossilisation ramp-ups determined by the required fleet exchange rates to achieve 100% defossilised transport by 2050 - the GHG emissions associated with the transport sector (including vehicle production and defossilised energy supply) will exceed the total 1.5 °C GHG budget (67% probability) for Europe (EU27+UK, all sectors) already by 2031/32. Since the cumulative GHG emissions 2021-2050 of all pathways are dominated by emissions of vehicle operation with fossil fuels (remaining vehicle fleet), a quick ramp-down of fossil fuel usage is most important to meet the climate targets. The ramp-up speed of completely sustainable transportation pathways, and thus the quickest ramp-down of fossil fuel usage, is "the crucial factor" for cumulative GHG emissions minimization. The faster carbon neutral energy can penetrate the existing market, the lower the cumulative GHG emissions.

Even with the assumed ramp-up of a new powertrain technology (passenger car 2033: 100% sales of new PT technology), keeping up the pace with the ramp-up of sustainable energy supply is a considerable challenge. Technically feasible ramp-ups of powertrain and energy/fuels supply are planned to be defined in a follow up study.
Defossilised drop-in fuels (carbon free on a WtW basis) are an option to eliminate GHG emissions of existing ICE powered vehicles and therefore could be an option to enable a faster introduction of GHG neutral energy supply to road transport. Therefore, significant efforts are required to defossilise gasoline and diesel fuel, which can be used in the existing vehicle fleet and non-electrifiable sectors. Electrolysis is a key element for a carbon neutral mobility sector. All pathways require significant electrolysis capacity, including BEV, since BEV in a fully renewable energy system requires a chemical buffer for dark doldrums. Ramp-up of electrolysis capacity is likely to become a temporary bottleneck.

Availability of critical raw materials is a key factor for enabling 100% BEV or 100% FCEV pathways. Potential bottlenecks have to be assessed in the global context. Lithium and cobalt demands are likely to become temporary bottlenecks in all 100% BEV scenarios with assumed battery technology assumptions (NMC 622, NMC 881 and solid-state Li-ion), 300 km / 500 km LDV range and the assumed future global transport increase in case of a full worldwide economic catch-up to the European Union until 2050. A mix of potential future battery technologies with lower specific Li and Co content, combined with slower increase of worldwide motorization and reduced battery sizes can reduce lithium and cobalt demand that does not exceed today known global resources. Furthermore, global lithium and cobalt resources and reserves have developed very dynamically in the last few years. Therefore, a considerable future increase of primary material supply can be expected.

Platinum demand will become a severe bottleneck in all 100% FCEV scenarios.
None of the investigated 100% pathways is restricted by technical-ecological land use bottlenecks or by other environmental impacts as eutrophication, PM formation and acidification.

Since ramp-ups of all pathways are likely to face temporary bottlenecks, a mix of technologies seems more robust to overcome those restrictions and is required to allow for quickest possible defossilisation and lowest cumulative GHG emissions.
While the analyzed theoretical 100% scenarios, where a single drivetrain technology and energy/fuel pathways is modelled to provide all of Europe's (road) transport demand allow for a comprehensive comparison of technologies, the most effective transformation will without doubt include a mix of technologies (which will be analyzed in a follow-up study). This expectation is supported by the results of this study, as depending on the applied metric, different technologies come out on top:

- With regard to minimum cumulative GHG emissions, identical ramp-up speeds would lead to very similar cumulative GHG emissions for hydrocarbon synthetic fuels, H2 (both for combustion engines as well as for fuel cells) and electric mobility. Any change of assumed ramp-up speed is likely to change the ranking of technologies.
- Regarding the lowest energy requirements, direct electrification (BEV) has the greatest advantage.
- Looking at total incremental costs, (short chain) synthetic hydrocarbon fuels, as methanol or methane are the least expensive options.

Considering the possibility of faster introduction than determined by the fleet exchange rate, legacy fleet compatible fuels, such as FT-gasoline/diesel could have a significant advantage, but only if the complete energy/fuel supply chain infrastructure (inclusive significantly large capacities of renewable power generation) can be built considerably quicker than the vehicle exchange rate allows. This is a major challenge, as the assumed scenarios already contain 28 % defossilised energy/fuel share in 2030.
Increasing vehicle efficiency is not always leading to an increase of overall efficiency. For FCEV and all ICE pathways e. g. light weight measures can increase the cumulative GHG emissions, if additional GHG from vehicle production outweighs GHG savings from efficiency improvements. Furthermore, the lowest total incremental costs, are achieved with state-of-the-art ICEV (no hybridization, no light-weight measures etc.) operated with synthetic fuels, since total costs of sustainable energy/fuel supply are lower than the vehicle on-costs for efficiency measures. Therefore, any efficient GHG avoidance policy requires a Life Cycle GHG reduction approach. If sector targets at set, they need to be well aligned with the life cycle approach.

7. Outlook

The follow-up study "FVV Fuels Study IVb" has already been started. First potential technical bottlenecks could be identified which are likely to decelerate the ramp-up of renewable mobility, sticking to exclusive 100%-scenarios, even under absolute ideal legal boundary conditions and investment attractiveness. These bottlenecks may involve the achievable ramp-up speed of key technologies as electrolysis, direct air capturing, reverse water gas shift reaction, charging infrastructure, electrical grid extension or build capacity of solar and wind power generation. To what extent these bottlenecks are limiting the required fast ramp-up of complete sustainable transportation pathways has not been finally investigated yet. Final results are expected by autumn 2022.

However, interim results turn more and more out to confirm, that concentration on a single energy-technology pathway is limiting the achievable ramp-up speed significantly. Most efficient GHG reduction requires intelligent technology mixes that enable the fasted possible exit out of fossil energy use for the lowest costs.

References

[1] FVV Fuels Study IV - Transformation of mobility to the GHG-neutral post-fossil age; FVV Final Report 1269, Frankfurt a.M., 2021; https://www.fvv-net.de/medien/aktuelles/detail/energiewende-im-verkehr-nur-mit-hilfe-der-modellierung-des-gesamten-energiesystems-kommt-man-zu-val/

[2] FVV Fuels Study III - Defossilizing the transportation sector; FVV Expert Paper (R586), Frankfurt a.M., 2018

Internationale Entwicklungen zu regenerativen Kraftstoffen

Franziska Müller-Langer, Karin Naumann, Jörg Schröder, Gabriel Costa de Paiva

Abstract

This presentation provides an overview on the development of renewable fuels worldwide. Also based on IEA studies an outlook will be given.

1. Hintergrund

Weltweit ist der langjährige Trend im Verkehrssektor durch einen hohen und stark wachsenden Energiebedarf gekennzeichnet. Wie bereits Bild 1 zeigt, hat sich in den vergangenen 30 Jahren seit 1990 dieser Bedarf nahezu verdoppelt. Auch wenn sich die Nutzung von Biokraftstoffen im gleichen Zeitraum auf etwa 3,8 EJ verfünffacht hat, so können damit nur ca. 3 % des Energiebedarfs durch eine biogene Alternative substituiert werden. Beim Strom ist der Zuwachs in den vergangenen 30 Jahren im Vergleich deutlich geringer ausgefallen (1,6-fach). Insgesamt decken diese beiden für die Energiewende wichtigen Energieträger gerade einmal 4,3 % des Energiebedarfs im globalen Verkehr ab.

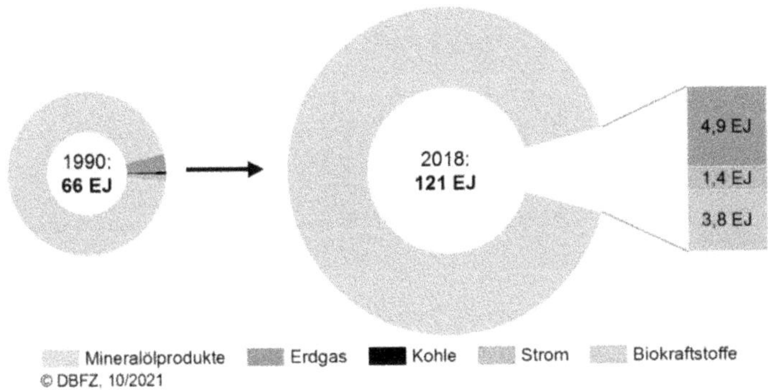

Bild 1: Weltweiter Endenergieverbrauch im Verkehrssektor nach Energiequellen in den Jahren 1990 und 2018

Treiber für die Nutzung erneuerbarer Kraftstoffe sind international teils sehr unterschiedlich. Eine anschauliche Übersicht über länderspezifische Entwicklungen unter unterschiedlichen Randbedingungen zeigt beispielsweise [2].

2. Produktion erneuerbarer Kraftstoffe aus Biomasse

Bereits im 20. Jahrhundert produzierte in Brasilien die zuckerrohrverarbeitende Industrie neben Zucker auch Bioethanol als Kraftstoff, in den 1990er-Jahren jährlich etwa 10 Mio. m^3, womit der Schwerpunkt der Welt-Bioethanolproduktion in Brasilien lag. Wie Bild 2 zeigt, stieg das weltweit produzierte Bioethanol auf 85 Mio. m^3/a (1,8 EJ/a) bis zum Jahr 2010 und anschließend bis zum Jahr 2019 auf etwa 110 Mio. m^3/a (2,3 EJ/a) an. Im Jahr 2020 ging die Produktion um etwa 10 % zurück, wohingegen für das Jahr 2021 wieder von einem leichten Wachstum ausgegangen wird. [IHS Markit (2021g)] Seit dem Jahr 2006 wird zudem ein relevanter Anteil der weltweit genutzten Biokraftstoffe in Form von Biodiesel (FAME) bereitgestellt. Nach einem kontinuierlichen Wachstum lag die produzierte Menge im Jahr 2019 bei etwa 40 Mio. t (1,5 EJ) FAME. Seit dem Jahr 2007 wird zudem auch HVO/HEFA (engl.: Hydrotreated Vegetable Oils bzw. Hydrotreated Esters and Fatty Acids) als paraffinisches Dieselsubstitut in großtechnischen Anlagen produziert. Diese Menge stieg bis zum Jahr 2019 auf etwa 6 Mio. t/a (0,3 EJ/a) weltweit.

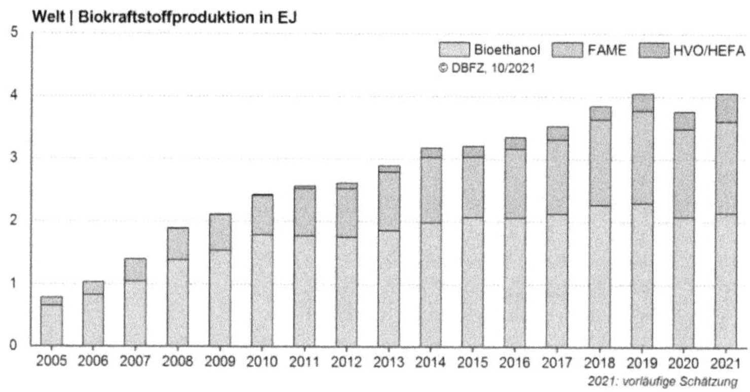

Bild 2: Entwicklung weltweiter Produktionsmengen an Biokraftstoffen

3. Nutzung und Handel

Die weltweite Nutzung von Biokraftstoffen entspricht weitestgehend der jährlichen Produktion, ab bzw. zuzüglich eventueller zwischenjähriger Verschiebungen durch Abnahme oder Aufbau von Lagerbeständen. Darüber hinaus findet eine regionale Verschiebung der Rohstoffe und Kraftstoffe zwischen den Erzeuger- und den Verbraucherländern statt.

In die Europäische Union wurden in den vergangenen Jahren mehr Biokraftstoffe importiert, als aus ihr exportiert wurden (Netto-Import). Der Netto-Import der beiden Biokraftstoffe mit den größten Marktanteilen ging von mehr als 1 Mio. m³ Bioethanol (2012) bzw. 2 Mio. t FAME (2011) auf 174.000 m³ Bioethanol (2018) bzw. 165.000 t FAME (2016) zurück. In den Folgejahren stieg der Netto-Import beider Kraftstoffe wieder (Bild 3 f.)

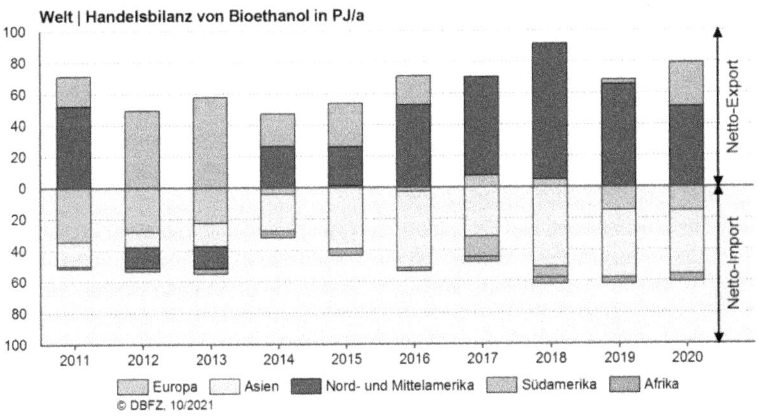

Bild 3: Handelsbilanz für Bioethanol

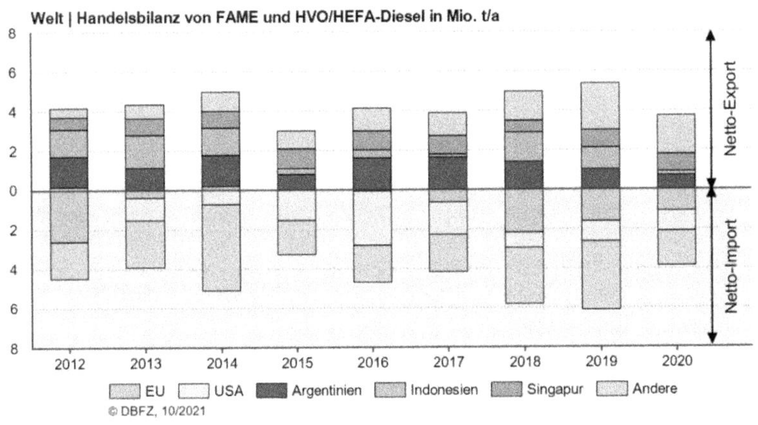

Bild 4: Handelsbilanz für FAME und HVO/HEFA

4. Ausblick

Eine anschauliche Analyse möglicher Entwicklungen des zukünftigen Mengenbedarfs für erneuerbare Kraftstoffe zeigt u.a. folgende Ergebnisse:
Dezidiert für den Straßenverkehr erfolgte dabei für ausgewählte Länder eine Untersuchung der Kraftstoffoptionen und alternativen Antriebe und ist mit mehreren Reports hier zu finden [4]. Zusammenfassend wurde daraus folgendes konstatiert:

- Erneuerbare Kraftstoffe können neben allen Formen von Elektrofahrzeugen einen wichtigen Beitrag zur Dekarbonisierung des Straßenverkehrssektors leisten, insbesondere kurz- und mittelfristig und für alle Verkehrsträger.
- Die Fälle für die Länderbewertung waren eine gute Auswahl, da die Zusammensetzung des Fahrzeugparks im Straßenverkehr und die Nutzungsraten der Energieträger so unterschiedlich sind
- Die derzeitige Politik führt zu einer Verringerung der TTW-CO_2-Emissionen, mit Ausnahme von Brasilien, aufgrund des erwarteten Wirtschaftswachstums >> Biokraftstoffe bieten die größten CO_2-Reduzierungen jetzt und bis 2030, 2040 und sogar 2050 für einige Länder
- Eine Maximierung des Einsatzes von Biokraftstoffen bringt mehr und schnellere CO_2-Reduzierungen. Selbst bei einer schnelleren Einführung von E-Fahrzeugen bleiben Biokraftstoffe kurz- bis mittelfristig der größte Beitrag zur Emissionsreduzierung.
- E-Kraftstoffe könnten genutzt werden, um die Lücke zwischen der verbleibenden Nachfrage nach fossilen Kraftstoffen und den Netto-Null-Zielen zu schließen. Die Menge an E-Fuels, die benötigt wird, um die verbleibenden fossilen Brennstoffe vollständig zu ersetzen, würde jedoch erhebliche Mengen an nichtfossilem Strom und abgeschiedenem CO_2 erfordern, die in vielen Ländern wahrscheinlich nicht verfügbar sind.
- Andere Dekarbonisierungsmaßnahmen wie Effizienzsteigerungen, elektrische Antriebe und Biokraftstoffe sind unerlässlich, um den Gesamtbedarf an E-Kraftstoffen zu senken.

Literatur

[1] Schröder, J.; Naumann, K. (Hrsg.) (2022): Monitoring erneuerbarer Energien im Verkehr. Leipzig: DBFZ. 340 S. ISBN: 978-3-946629-82-5. DOI: 10.48480/19nz-0322.

[2] IEA Bioenergy TCP, Task 39 (2022): Implementation Agendas: Compare-and-Contrast Transport Biofuels Policies (2019-2021 Update) ISBN# 979-12-80907-06-6

[3] IEA (2021): Net Zero by 2050 - A Roadmap for the Global Energy Sector; https://iea.blob.core.windows.net/assets/deebef5d-0c34-4539-9d0c-10b13d840027/NetZeroby2050-ARoadmapfortheGlobalEnergySector_CORR.pdf

[4] IEA AMF, IEA TCP Bioenergy, EC (2020): The Role of Renewable Fuels in Decarbonizing Road Transport, https://iea-amf.org/content/projects/map_projects/58

Neue Motor-, Abgas-, und Kraftstofftechnologien – Auswirkungen auf die Dieselmotoremissionen und die Gesundheit

Jürgen Bünger, Axel Munack, Jürgen Krahl

Abstract

Populations around the world including workers at many workplaces are exposed to diesel engine emissions (DEE). DEE can contribute to various respiratory, pulmonary and cardiovascular health risks. During the last decades, many researchers – including the Fuels Joint Research Group (FJRG) – conducted hundreds of studies on toxicological characteristics of DEE from biogenic and petrol diesel fuels and their biological effects using analytical chemistry, biological short-term assays, as well as animal experiments and human studies.
Driven by strict exhaust gas regulations around the world, many studies have confirmed the reduction of DEE – especially particles and nitrogen oxides – from individual modern diesel engines as well as overall in nearly all environments. In summary, convincing decreases of toxicants in diesel engine emissions were confirmed over time using new engine-, exhaust- and fuel-technologies.
This review aims at an evaluation, how these undoubted improvements affect the burdens and possible health risks of people? Are there still health risks from actual emission levels? A literature search of recent scientific publications was conducted to answer this question. Different organs (endpoints) were included in the search like the airways, the cardiovascular system, as well as neurological and immunological processes and cancer. In addition, a distinction was made between acute and chronic effects of the exhaust.

1. Einleitung

Weltweit sind ein Großteil der Allgemeinbevölkerung und der Beschäftigten am Arbeitsplatz durch Dieselmotoremissionen (DME) exponiert. DME können zu verschiedenen Gesundheitsrisiken der Atemwege, des Herz-Kreislauf-Systems und anderer Organe beitragen. Aus arbeitsmedizinisch-epidemiologischen Studien geht hervor, dass akute und chronische Effekte von DME vor allem mit hohen Partikelexpositionen assoziiert sind. Im Jahr 2012 wurden DME von der International Agency for Research on Cancer (IARC), einer Tochterorganisation der WHO, als mutagen und humankanzerogen eingestuft [1]. Außerdem sind Dieselabgase positiv (begrenzte Evidenz) mit einem erhöhten Blasenkrebsrisiko assoziiert.
Allerdings wurde in einer umfassenden Bewertung des Diesel Epidemiology Panels des US-amerikanischen Health Effects Institutes 2015 festgestellt, dass zukünftige Ri-

sikobewertungen auch die großen zwischenzeitlichen Änderungen bei Dieselkraftstoffen, Motoren und Nachbehandlungstechnologien berücksichtigen müssen, seit diese Studien durchgeführt wurden, da sie starke Auswirkungen auf die Umweltkonzentrationen der Schadstoffe haben [2]. Zu einer ähnlichen Bewertung kommt Claxton in einem umfassenden Review und fordert zusätzliche Forschung und bessere Expositionsbewertungen, damit Politiker und die Öffentlichkeit entscheiden können, ob Dieselmotoren durch andere Technologien ersetzt werden sollten [3].

2. Ergebnisse

In den toxikologischen und medizinischen Datenbanken und der Webseite des Umweltbundesamtes wurde eine Literaturrecherche zwischenzeitlich publizierter wissenschaftlicher Studien und Übersichtsarbeiten zur Reduktion der Emissionen und deren Einflüsse auf die möglichen gesundheitlichen Risiken durchgeführt, die zusammen mit eigenen Daten der Fuel Joint Research Group (FJRG) dargestellt werden.

2.1 Emissionen

Dieselmotoren neuer Technologie, betrieben mit schwefelarmem Dieselkraftstoff und Abgasnachbehandlung durch einen DPF, haben die PM-Emissionen in den U.S.A. gegenüber älteren Motoren um ca. 99% reduziert. Krebserregende Emissionen von PAK und Nitro-PAK sind im Vergleich zum Jahr 2004 um 80% - 99% gesunken [4, 5].
Auch in Europa haben die Emissionen in vergleichbarem Ausmaß abgenommen. Im deutschen interdisziplinären Forschungsverbund aus Ingenieuren, Chemikern und Ärzten (Fuel Joint Research Group, FJRG) wurden DME und ihre biologischen Effekte untersucht [6]. Immer wurden auch neue biogene und synthetische Kraftstoffe im Vergleich zu verschiedenen Mineralöldieselqualitäten getestet. Aus 32 Projekten wurden die Daten der DME zu gesetzlich limitierten Komponenten zusammengestellt und im Zeitverlauf analysiert (Abb.1).

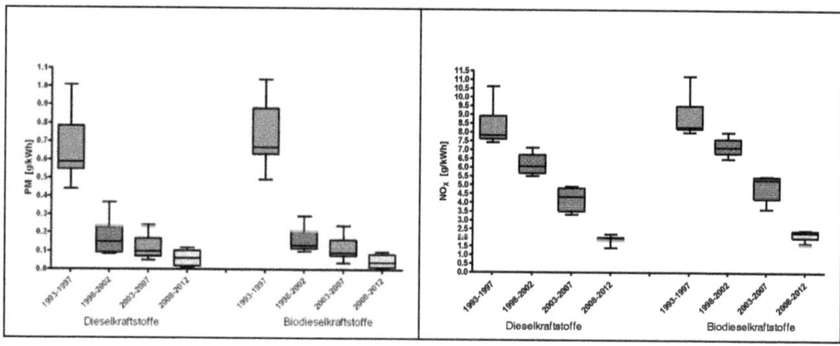

Abb. 1: Reduktion der Partikelmasse (PM, links) und der Stickoxide (NO_X, rechts) im Dieselrohabgas.

Die Reduktionen im Dieselrohabgas gehen auf die verbesserte Motortechnologie und die Neuentwicklung bzw. Reformulierung von Kraftstoffen zurück. Mit Dieselpartikelfilter (DPF) reduzieren sich die Partikelmassen weiter um bis zu 99% und die NO_X mit SCR-System um 50 - 90% [6, 7, 8].

Dass diese stark reduzierten Emissionen der einzelnen Motoren bzw. Fahrzeuge auch die Umwelt entlastet haben, zeigen die Ergebnisse für die spezifischen Pkw- und Lkw-Emissionen (Abb. 2 und 3) in der Emissionsberichterstattung des Umweltbundesamtes 2020 im Modell TREMOD (Transport Emission Modelling) im Zeitverlauf von 1995 bis 2020. Dass die DME am stärksten abgesenkt wurden, zeigt der Vergleich zwischen Pkw (Abb. 2) und Lkw (Abb. 3), da die Lkw-Flotte fast ausschließlich von Dieselmotoren angetrieben wird und hier die Absenkung noch deutlicher wird als bei den Pkw.

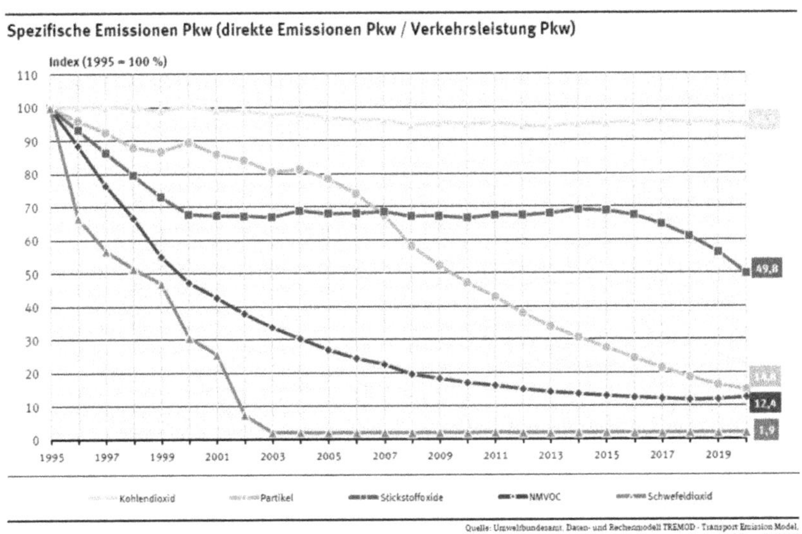

Abb. 2: Reduktion spezifischer Pkw-Emissionen von 1995 bis 2020.

Für die deutliche Reduktion der Emissionen spricht auch, dass es 2021 in Deutschland erneut keine Überschreitungen der Feinstaubgrenzwerte gab. Der Jahresmittelgrenzwert für Stickstoffdioxid (NO_2) von 40 µg/m³ Luft wurde voraussichtlich nur noch an ein bis zwei Prozent der verkehrsnahen Messstationen überschritten. Das zeigt die vorläufige Auswertung der Messdaten der Länder und des Umweltbundesamtes (Stand 31.01.2022) von bislang rund 600 Messstationen [9].

Medizinisch-toxikologisch ist gesichert, dass den Partikeln in den DME die gesundheitlich weitaus größte Bedeutung zukommt. Die Stickoxide werden als weniger relevant eingestuft. Ihre Wirkung ist separat auch kaum darstellbar, da die Emission hochgradig korreliert ist zu den Partikeln, wie aus Abb. 4 sehr gut zu entnehmen ist.

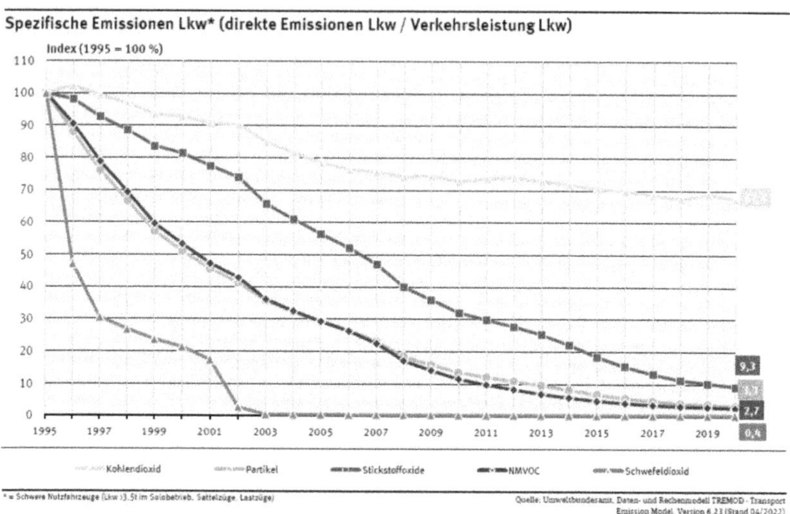

Abb. 3: Reduktion spezifischer Lkw-Emissionen von 1995 bis 2020.

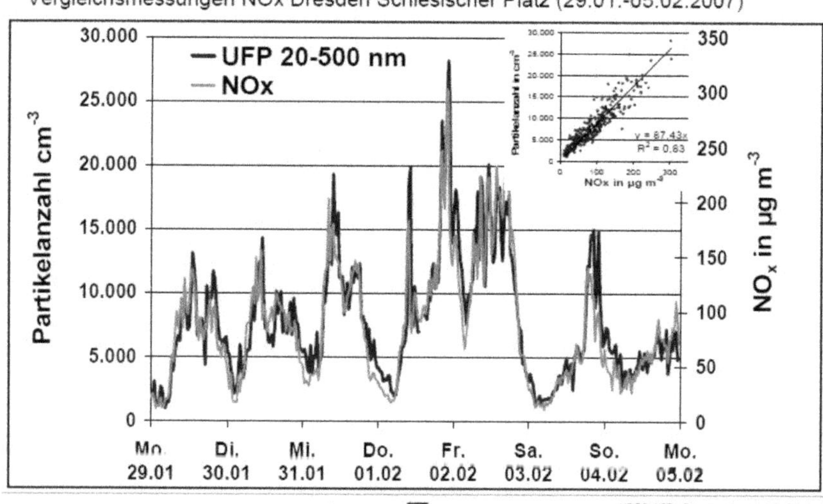

Abb. 4: Korrelation zwischen ultrafeinen Partikelemissionen (UFP) und Stickoxiden in der Umwelt.

2.2 Wirkungen

DME können auf verschiedene Organsysteme Einfluss nehmen. Im Vordergrund stehen die Wirkungen auf die Lunge, den primären Ort der Einwirkung nach Inhalation. Mögliche chronische Effekte wie chronisch-obstruktive Bronchitis (COPD) und Lungenkrebs sind durch Entzündungsreaktionen auf die eingeatmeten Partikel bedingt. Blasenkrebs wird am ehesten durch die PAK und Nitro-PAK im Abgas verursacht. Weitere Effekte sind weniger gut belegt wie zum Beispiel auf Herz und Kreislauf, Nervensystem oder das Immunsystem. Da die Wirkungen der Partikel entscheidend für die möglichen gesundheitlichen Wirkungen sind, wurden sie am besten untersucht und sind Gegenstand der meisten relevanten Studien und Übersichtsarbeiten. Im Weiteren wird daher überwiegend auf die Wirkungen von Partikelemissionen eingegangen. Akute Effekte werden im medizinischen Alltag sehr selten beobachtet, werden aber experimentell *in vitro*, im Tierversuch und auch an Menschen untersucht, um Hinweise für eventuelle Langzeitrisiken zu erhalten. Die Humanstudien werden wegen ihrer besonderen Bedeutung und besseren Übertragbarkeit auf die reale Alltagssituation bevorzugt dargestellt.

Akute Effekte:
Im Jahr 2020 analysierten Weitekamp und Koautoren [10] 26 experimentelle Studien an Tieren und Menschen mit Expositionen durch Gesamt-DME und durch mit Partikelfiltern gereinigte DME. Die Höhe der Partikelexposition betrug in den Gesamt-DME zwischen 35,5 und 2.500 µg/m^3 Luft und für das gefilterte Abgas (22 Studien) 0 bis 40 µg/m^3. In 4 Tierstudien ergaben sich erhöhte Werte für Zytokine, Makrophagen und neutrophile Granulozyten in der Lungenspülflüssigkeit, aber keine sonstigen Auffälligkeiten im Bereich der Atemwege. Herz-Kreislauf-Effekte wurden in 3 Tierstudien untersucht und ergaben sehr widersprüchliche Resultate. Zu Auswirkungen auf das Nervensystem wurden 3 Tierstudien aus einer Forschergruppe bewertet, die im Hinblick auf den Menschen wenig aussagekräftig sind. In 4 Tierstudien wurden Auswirkungen auf das Immunsystem untersucht. Nach Inhalation von DME wurden vermehrt Zytokin- und Chemokin-Erhöhungen gemessen, ohne dass daraus eine Bedeutung für den Menschen ableitbar ist [10].
In einer randomisierten, doppelblinden Crossover-Studie wurden 14 Allergen-sensibilisierte Personen (9 mit Überempfindlichkeit der Atemwege) einer inhalativen Allergenprovokation unterzogen, nachdem sie 2 Stunden lang unbehandelten DME, gefilterten DME oder Luft ausgesetzt waren [11]. Die Exposition durch DME und Allergen erhöhte die Zahl der neutrophilen Leukozyten im Blut und war nach 48 Stunden mit anhaltender Eosinophilie verbunden. Veränderungen der peripheren Leukozyten korrelierten mit einer Abnahme der Lungenfunktion.
Wirkungen auf das Herz-Kreislaufsystem wurden in 2 Humanstudien – darunter auch an Patienten mit Herzerkrankungen untersucht. Die Ergebnisse hinsichtlich einer Wirkung auf die Erweiterung bzw. Verengung von Arterien waren widersprüchlich [10]. In einer weiteren Studie wurde der Einsatz eines Partikelfilters auf kardiovaskuläre Wirkungen einer Abgasinhalation bei Männern untersucht [12]. Die Autoren beobachteten

positive Auswirkungen auf Biomarker der kardiovaskulären Gesundheit und kamen zu dem Schluss, dass der Einsatz von Partikelfiltern in dieselbetriebenen Fahrzeugen erhebliche Vorteile für die öffentliche Gesundheit hat und die Belastung durch Herz-Kreislauf-Erkrankungen verringert.
Trotz bislang noch begrenzter Tests am Menschen scheint auch eine kurzfristige Exposition gegenüber Biodieselabgasen nur mit kardiopulmonalen Effekten verbunden zu sein, die mit denen von herkömmlichen DME vergleichbar sind [13, 14].

Chronische Effekte:
Der Einfluss von DME auf chronische Atemwegserkrankungen wie COPD ist immer wieder beschrieben worden. Allerdings sind dazu offensichtlich sehr hohe Expositionen erforderlich. In einer aktuellen Studie [15] wurden DME-exponierte Bergleute im Kali- und Salzbergbau untersucht und keine negativen Einflüsse auf die Lungenfunktion beobachtet trotz deutlicher Überschreitungen des Arbeitsplatzgrenzwertes für DME von 0,05 mg/m³ (gemessen als Elemental Carbon in der alveolengängigen Fraktion). Auch Auswirkungen auf das Herz-Kreislauf-System wurden untersucht, indem sonographisch die Dicke der Innenschichten der Halsarterie (Carotis-Intima-Media-Thickness, CIMT) als Biomarker der Arteriosklerose gemessen wurde. Auch hier wurden keine Hinweise für eine vermehrte Arteriosklerose der DME-exponierten Bergleute im Vergleich zu nicht bzw. gering exponierten Beschäftigten über Tage gefunden. Blutuntersuchungen auf Entzündungsmarker verliefen ebenfalls weitestgehend unauffällig [15]. Für DME aus der Verbrennung biogener Kraftstoffe liegen keine entsprechenden Studien vor.
Im Tierversuch führte eine DME-Exposition aus moderner Technologie nicht zu Veränderungen der Plasmamarker für eine allgemeine Toxizität oder von Veränderungen an kardiovaskulären Organen. Hinweise auf DME-induzierte Wirkungen an weiblichen Ratten wurden dem durch Alterungsprozesse der Tiere bedingten Östrogenmangel im Verlauf der 2jährigen Studie zugeordnet [16].

Zu den chronischen Wirkungen der DME gehören auch die mutagenen und krebserregenden Eigenschaften, die durch die PAK-beladenen Partikel verursacht werden und zu Lungenkrebs führen können. Da die Entstehung von Krebs – ähnlich wie beim Rauchen – aber in der Regel 30 Jahre und mehr beträgt, kann sie nur in Langzeitstudien zuverlässig untersucht werden. Diese liegen für DME aus moderner Technologie naturgemäß (noch) nicht vor. Hilfsweise werden daher Tierstudien und Tests an Zellkulturen (*in vitro*) zur Bewertung herangezogen.
McDonald und Koautoren untersuchten Effekte einer 2jährigen Exposition von Nagetieren gegenüber Emissionen aus Motoren neuerer Technologie ab 2007 und fanden keine krebsverdächtigen Effekte [17].
Auch für *in vitro*-Studien forderte Larry Claxton [3]: „It would be to industry's advantage to demonstrate that both chemical and bioassay indicators of genotoxicity and carcinogenicity are dramatically reduced in modern vehicles." Dies konnte zwischenzeitlich sowohl für Mineralöl-basierte als auch für biogene Dieselkraftstoffe gezeigt werden

[18, 19, 20]. Die von der FJRG für DME in den ersten Studien beobachtete hohe Mutagenität sank über zwei Jahrzehnte kontinuierlich ab (Abb. 5) [7, 8, 13]. Bei Pkw mit Euro 5- und Euro 6-Zertifizierung und einem Lkw mit Euro VI-Standard war die Mutagenität in dem durchgehend seit 1993 verwendeten Standard-Protokoll des bakteriellen Rückmutationstestes gegenüber der Spontanmutationsrate nicht mehr signifikant erhöht [21].

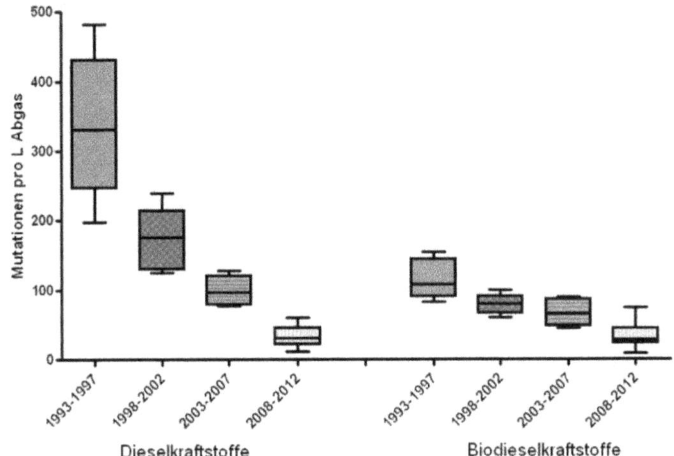

Abb. 5: Reduktion der mutagenen Effekte von Dieselabgas im Zeitverlauf.

In einigen Kraftstoffprojekten wurden allerdings unerwartet hohe mutagene Effekte beobachtet, obwohl die gesetzlich limitierten Emissionen für diese Kraftstoffe keine oder nur geringe Auffälligkeiten ergaben. So wiesen die Abgaspartikulate von Rapsöl in einem modernen Serien-Lkw-Motor eine stark erhöhte Mutagenität auf [22]. Mischkraftstoffe (80% DK, 20% RME) führten ebenfalls zu einem Anstieg der mutagenen Effekte der Partikelemissionen, deren Ursache offensichtlich die Bildung von Oligomeren aus den Kraftstoffkomponenten ist [20, 23].

3. Fazit und Ausblick

In den letzten Jahren wurden DME aus Pkw und Lkw und das damit verbundene Gesundheitsrisiko insgesamt bedeutend abgesenkt. Da nach dem derzeitigen wissenschaftlichen Erkenntnisstand die Partikel die wesentlichen Effekte der DME verursachen, bedeutet deren Reduktion mit hoher Wahrscheinlichkeit, dass auch die damit verbundenen gesundheitlichen Risiken entsprechend abgesenkt wurden. Selbst in Arbeitsbereichen, wo diese Absenkungen noch nicht im vollen Umfang umgesetzt werden konnten, z. B. beim Betrieb von Dieselmotoren in Gebäuden und unter Tage, konnten keine erhöhten Gesundheitsrisiken für die Beschäftigten mehr festgestellt werden. Ob auf dem derzeit erreichten niedrigen Niveau noch ein erhöhtes Risiko durch DME für die Allgemeinbevölkerung besteht, ist nach der vorliegenden Datenlage unwahrscheinlich.

Wie einige Studien der FJRG zeigen, müssen trotzdem zur Risikobewertung und im Sinne der Prävention neue Technologien und Kraftstoffe durch Emissionsmessungen und Kurzzeit-Screening-Tests hinsichtlich der zu erwartenden Emissionen und Risiken untersucht werden.

Literatur

[1] Benbrahim-Tallaa L, Baan RA, Grosse Y, Lauby-Secretan B, El Ghissassi F, Bouvard V, Guha N, Loomis D, Straif K, International Agency for Research on Cancer Monograph Working Group (2012) Carcinogenicity of diesel-engine and gasoline-engine exhausts and some nitroarenes. Lancet Oncol. 13:663-664.

[2] HEI Diesel Epidemiology Panel (2015) Diesel Emissions and Lung Cancer: An Evaluation of Recent Epidemiological Evidence for Quantitative Risk Assessment. Res Rep Health Eff Inst. 19:1-149.

[3] Claxton LD (2015) The history, genotoxicity, and carcinogenicity of carbon-based fuels and their emissions. Part 3: diesel and gasoline. Mutat Res Rev Mutat Res. 763:30-85.

[4] Khalek IA , Bougher TL, Merritt PM, Zielinska B (2011) Regulated and unregulated emissions from highway heavy-duty diesel engines complying with U.S. Environmental Protection Agency 2007 emissions standards J Air Waste Manag Assoc. 61:427-442.

[5] Khalek IA, Blanks MG, Merritt PM, Zielinska B (2015) Regulated and unregulated emissions from modern 2010 emissions-compliant heavy-duty on-highway diesel engines. J Air Waste Manag Assoc. 65:987-1001.

[6] Schröder O, Bünger J, Munack A, Knothe G, Krahl J (2013) Exhaust emissions and mutagenic effects of diesel fuel, biodiesel and biodiesel blends. Fuel 103, 414-420.

[7] Bünger J, Munack A, Schröder O, Krahl J (2013) Mutagenic effects of emissions from old and new technology diesel engines combusting biogenic and fossil fuel qualities. 1st International Conference and 6th International Workshop of the Cluster of Excellence "Tailor-Made Fuels from Biomass", Aachen, S. 12-13.

[8] Bünger J, Westphal G, Brüning T, Schröder O, Munack A, Krahl J (2013) Dieselmotoremissionen und ihre biologischen Effekte - Ein Überblick aus 20 Jahren eigener Studien. In: Krahl, J., Munack, A., Eilts, P., Bünger, J. (Hrsg.) in: Kraftstoffe von heute und morgen: 6. Biokraftstoffsymposium am 27.+ 28.02.2014 in Coburg. Fuels Joint Research Group Band 10, Cuvillier-Verlag, Göttingen, 2015.

[9] UBA (2022) Umweltbundesamt, Luftqualität 2021, Vorläufige Auswertung, S. 1-34, www.umweltbundesamt.de/publikationen, Kurzlink: bit.ly/2dowYYI.

[10] Weitekamp CA, Kerr LB, Dishaw L, Nichols J, Lein M, Stewart MJ (2020) A systematic review of the health effects associated with the inhalation of particle-filtered and whole diesel exhaust. Inhal Toxicol. 32:1-13.

[11] Wooding DJ, Ryu MH, Hüls A, Lee AD, Lin DTS, Rider CF, Yuen ACY, Carlsten C (2019) Particle Depletion Does Not Remediate Acute Effects of Traffic-related Air Pollution and Allergen. A Randomized, Double-Blind Crossover Study. Am J Respir Crit Care Med. 200:565-574.

[12] Lucking AJ, Lundbäck M, Barath SL, Mills NL, Sidhu MK, Langrish JP, Boon NA, Pourazar J, Badimon JJ, Gerlofs-Nijland ME, Cassee FR, Boman C, Donaldson K, Sandstrom T, Newby DE, Blomberg A (2011) Particle traps prevent adverse vascular and prothrombotic effects of diesel engine exhaust inhalation in men. Circulation. 123:1721-1728.

[13] Bünger J, Krahl J, Munack A, Brüning T, Westphal G (2014) Wie riskant sind Biokraftstoffe? Naturwissenschaftliche Rundschau 7 (67), S.332-342.

[14] Godri Pollitt KJ, Chhan D, Rais K, Pan K, Wallace JS (2019) Biodiesel fuels: A greener diesel? A review from a health perspective. Sci Total Environ. 688:1036-1055.

[15] Gamrad-Streubel L, Haase LM, Rudolph KK, Rühle K, Bachand AM, Crawford L,. Mundt KA, Bünger J, Pallapies D,Taeger D, Casjens S, Molkenthin A, Neumann S, Giesen J, Neumann V, Brüning T, Birk T (2022) Underground Salt and Potash Workers Exposed to Nitrogen Oxides and Diesel Exhaust – Assessment of Specific Effect Biomarkers". International Archives of Occupational and Environmental Health, in press.

[16] Conklin DJ, Kong M (2015) Advanced Collaborative Emissions Study (ACES): Lifetime Cancer and Non-Cancer Assessment in Rats Exposed to New-Technology Diesel Exhaust, Part 4. Assessment of plasma markers and cardiovascular responses in rats after chronic exposure to new-technology diesel exhaust in the ACES bioassay. HEI Health Review Committee. Res Rep Health Eff Inst. 184:107-139.

[17] McDonald JD, Doyle-Eisele M, Seagrave J, Gigliotti AP, Chow J, Zielinska B, Mauderly JL, Seilkop SK, Miller RA (2015) Advanced Collaborative Emissions Study (ACES): Lifetime Cancer and Non-Cancer Assessment in Rats Exposed to New-Technology Diesel Exhaust, Part 1: Assessment of carcinogenicity and biologic responses in rats after lifetime inhalation of new-technology diesel exhaust in the ACES bioassay. HEI Health Review Committee. Res Rep Health Eff Inst. 184:9-44.

[18] Claxton LD (2015) The history, genotoxicity and carcinogenicity of carbon-based fuels and their emissions: part 4 - alternative fuels. Mutat Res Rev Mutat Res. 763:86-102.

[19] Bünger J, Krahl J, Schröder O, Schmidt L, Westphal GA (2012) Potential hazards associated with combustion of bio-derived versus petroleum-derived diesel fuel. Crit Rev Toxicol. 42:732-750.

[20] Bünger, J (2016) Gentoxizität von Dieselmotoremissionen bei Verbrennung von Pflanzenölen, Mineralöldiesel und deren Mischkraftstoffen. In: Fuels Joint Research Group, Band 17, (Hrsg.: Krahl J, Munack A, Eilts P, Bünger J), Cuvillier-Verlag, Göttingen, 72 S.

[21] Götz K, Fey B, Singer A, Krahl J, Bünger J, Knorr M, Schröder O (2016) Exhaust gas emissions and engine oil interactions from a new bio-based fuel named Diesel R33. SAE International, 2016-01-2256.

[22] Bünger J, Krahl J, Munack A, Ruschel Y, Schröder O, Emmert B, Westphal G, Müller M, Hallier E, Brüning T (2007) Strong mutagenic effects of diesel engine emissions using vegetable oil as fuel. Arch Toxicol. 81, 599-603.

[23] Bünger J, Krahl J, Munack A, Ruschel Y, Schröder O, Handrich C, Müller M, Hallier E, Brüning T (2009) Erhöhte Gentoxizität durch Dieselmotoremissionen bei Verbrennung von Kraftstoffmischungen mit Biodieselanteil; in: 49. Jahrestagung der Deutschen Gesellschaft für Arbeitsmedizin und Umweltmedizin, S. 237 – 240.

Alternative Kraftstoffe im Premium Automobilbereich

Hanno Krämer, Markus Send

Abstract

During 4th FJRG 2021 conference, Audi presented mixture experiments towards a potential high-octane, high-quality drop-in reFuel within the limits of EN 228 E5. Finally, a fuel formulation could be found, mainly based on iso-butene derivatives. In this study this fuel -now called Audi MTO- is compared to some well-known fossil fuel types representing the whole bandwidth from very clean burning to very sooting candidates on a dynamic engine test bench regarding gaseous and particulate emissions during WLTC cycles. The Audi MTO reFuel type shows remarkable low PN emissions at standard start temperature of 23 °C as well as at a cold start temperature of 0 °C, while the gaseous emissions remain in the typical range of all other fuels. Thus, fuels of type Audi MTO can simultaneously help reduce CO_2 emissions by using regenerative production paths as well as reduce environmental impacts of even older vehicles by very clean burning behavior.

1. Einleitung

Insbesondere im Premium-Automobilbereich ist mit erhöhten Lebens- und Haltedauern der Fahrzeuge zu rechnen. Alternative Kraftstoffe, die dieses Fahrzeugsegment bedienen sollen, müssen daher nicht nur für zukünftige Fahrzeugentwicklungen geeignet sein, sondern besonders auch von der bestehenden Fahrzeugflotte nutzbar sein. Dies bedingt für die Kraftstoffe möglichst eine Erfüllung der DIN EN 228 und hier im besten Fall die E5-Variante mit max. 2,7% Sauerstoffgehalt. Im Rahmen der 4. Tagung der Fuels Joint Research Group Tagung 2021 wurde ein solcher Kraftstoff von Zusammensetzung und Siedeeigenschaften her seitens Audi vorgestellt [1]. Sowohl Siedeverlauf als auch chemische Zusammensetzung weckten die Hoffnung, hier einen Kraftstoff mit äußerst sauberen Verbrennungseigenschaften vorliegen zu haben. Offen blieb zum damaligen Zeitpunkt jedoch, wie sich dieser Vorschlag hinsichtlich Emissionen im Vergleich zu klassischen Kraftstoffen unterschiedlichen Zuschnitts, und ggf. zu einem Idealkraftstoff im motorischen Betrieb tatsächlich verhält. Dieser Fragestellung soll im Folgenden nachgegangen werden.

2. Testmethode und Versuchsbedingungen

Zum Einsatz kam ein 4-Zylinder-Motor der Baureihe EA888 von Audi, der auf einem dynamischen Prüfstand im WLTC bei zwei verschiedenen Starttemperaturen betrieben wurde. Die Eckdaten des Aggregates sind in Tabelle1 zusammengefasst.

Tabelle1: Motordaten

Motor:	2.0l TFSI EA888 evo4 MLB LK2 EU6AP
Typ:	4 Zylinder, Reihe
Bohrung x Hub:	82,50 mm x 92,80 mm
Hubvolumen:	1984 ccm
Einspritzsystem:	Direkteinspritzung, seitliche Injektorlage
Einspritzdruck:	350 bar
Max. Leistung / 1/min:	180 kW / 5000 – 6500
Max. Moment / 1/min:	380 Nm / 1600 – 4300
Getriebe:	DL382 7-Gang S-Tronic
Schwungmassenklasse:	2268 kg

Vermessen wurden sowohl die gasförmigen Tailpipe-Emissionen als auch die Roh-Emissionen der Partikelanzahl. Als Vergleichskraftstoffe dienten der gesetzliche Zertifizierungskraftstoff für EU5, welcher sich in der Vergangenheit als zwar insgesamt hochwertiger, aber stark partikelbildender Kraftstoff erwiesen hat, tankstellenähnliche Ware der Qualität 95 E10 und als Idealkraftstoff der Erstbefüllkraftstoff „ASF" des VW-Konzerns. Einige wichtige Eckdaten der verwendeten Kraftstoffe sind in Tabelle 2 zusammengestellt. Hierbei bedeuten die grün hinterlegten Felder günstige Kraftstoffeigenschaften hinsichtlich möglichst niedriger PN-Emissionen, während die gelben Felder als kritisch zu betrachten sind.

Tabelle2: Eigenschaften der Kraftstoffe

		ASF	MTO Audi	RON95 E10	EU5cert Audi
RON	-	96	107	96	101
DVPE	kPa	-	68	58	59
Density	kg/m3	690	722	749	751
EVAP@150	%	100	78	94	84
T90	°C	105	174	143	175
FBP	°C	147	184	170	204
Aromatics	%v	0	0	31	32
Aromatics >C8	%v	0	0	7	16
YSI_calc	-	46	62	97	116
Oxygen	%m	2,5	2,7	3,5	1,8
Ether	%v	14	15	0	0
Ethanol	%v	0	0	9	5
LHV (mass.)	MJ/kg	43,5	43,2	41,9	42,0
LHV (vol.)	MJ/L	30,0	31,2	31,4	31,5

3. Ergebnisse

Bild 1 zeigt die gasförmigen Emissionen NO_x, HC und CO der genannten Kraftstoffe bei zwei unterschiedlichen Starttemperaturen im Vergleich zu den gesetzlichen Grenzen von EU6 AP. Die Messung entspricht nicht exakt einer Homologationsmessung, da hier ein dynamischer Motorprüfstand anstelle eines konditionierten Fahrzeuges auf der Rolle eingesetzt wurde. Dennoch erhält man einen guten Überblick, wie sich die unterschiedlichen Kraftstoffe auf das Emissionsverhalten auswirken. Bei den gasförmigen Emissionen ist kein systematischer Einfluss der Kraftstoffe zu erkennen. Alle Emissionen befinden sich weit unterhalb des gesetzlichen Limits, was angesichts eines neuen, nicht gealterten Katalysators auch erwartet werden darf. Eindeutig zu erkennen ist dagegen der Einfluss der Starttemperatur, die 0 °C-Startbedingung führt regelmäßig zu höheren Emissionen. Dieser Effekt ist bei den NO_x-Emissionen etwas schwächer ausgeprägt als bei CO und insbesondere HC.

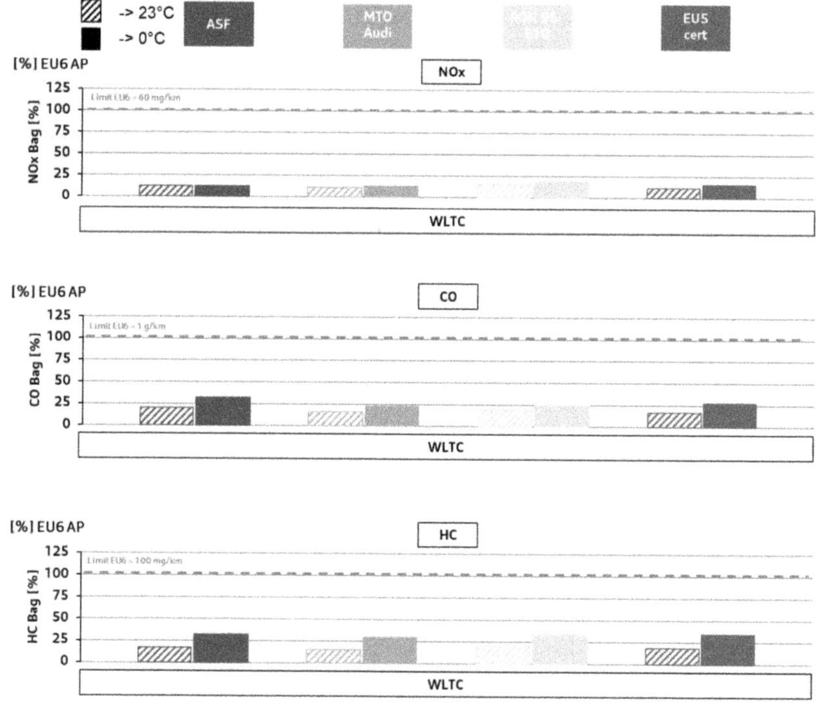

Bild 1: Vergleich der gasförmigen Emissionen der Kraftstoffe

Bild 2 zeigt die Roh-Emissionen des Motors hinsichtlich PN. Die Kraftstoffe sind von der Reihenfolge her entsprechend ihrem berechneten Yield Sooting Index (YSI) angeordnet, beginnend links mit dem niedrigsten YSI. Es ist eindeutig zu erkennen, dass der YSI eine gute Vorhersagegröße für die resultierenden PN-Emissionen darstellt, sowohl bei der Starttemperatur von 23 °C als auch bei 0 °C. Aus der Starttemperatur von 0 °C resultiert für alle Kraftstoffe ein um einen Faktor von ca. 4-6 höheres Partikelniveau als bei der wärmeren Starttemperatur.

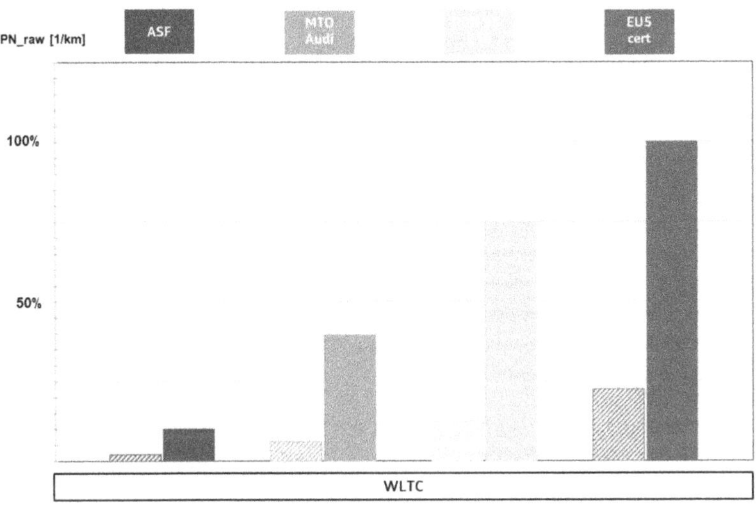

Bild 2: Vergleich der PN-Emissionen der Kraftstoffe

Im Vergleich der Kraftstoffe zeigt sich der ASF als ein extrem sauber verbrennender Kraftstoff, während auf der anderen Seite der EU5-Kraftstoff teilweise mehr als 10fach höhere PN-Emissionen erzeugt. Der Audi-MTO-Kraftstoff erweist sich bei warmen Bedingungen als sehr nahe an der ASF-Qualität befindlich, büßt allerdings einen Teil dieses Vorteils bei Kälte aufgrund seines deutlich höheren T90-Punktes von 174 °C und seines ebenfalls höheren Siedeendes von 184 °C im Vergleich zum ASF wieder ein. Beides bewirkt eine etwas schlechtere Gemischbildung im kalten Motor mit einem evtl. auch leicht erhöhten Risiko von Wandbenetzung, was jeweils dazu beitragen kann, die Partikelemissionen der Verbrennung zu erhöhen. Dennoch bleibt Audi-MTO bei beiden Starttemperaturen weit unterhalb des PN-Emissionsniveaus von sowohl der tankstellenähnlichen ROZ95 E10-Ware als auch mit noch größerem Abstand zum EU5-Kraftstoff. Zwischen den beiden letztgenannten Kraftstoffen ist sehr gut zu erkennen, welch erhebliche Auswirkung eine ungünstige Kraftstoffzusammensetzung selbst innerhalb der EN228 hinsichtlich der PN-Emissionen hat. Insgesamt bestätigt sich, dass es durchaus möglich ist, nicht nur eine hochoktanige, aromatenfreie Kraftstoffrezeptur innerhalb der EN228 E5 darzustellen, sondern dass diese in unserem Fall darüber hinaus auch mit sehr niedrigen PN-Emissionen einhergeht.

4. Zusammenfassung

Alternative, aromatenfreie Kraftstoffe mit abgesenktem Siedeende haben großes Nutzpotential im motorischen Betrieb. Was sich aus den Stoffdaten schon ablesen ließ, hat sich in der Anwendung am Motorprüfstand dann auch bestätigt. Insbesondere die Partikelanzahlemissionen (PN) lassen sich hierdurch im Vergleich selbst zu einer guten Tankstellenware nahezu halbieren, der Vergleich zu ungünstig zusammengesetztem Kraftstoff fällt entsprechend noch viel deutlicher aus. Profitieren kann hiervon der gesamte Fahrzeugbestand, da sich der hier als Beispiel gezeigte Audi-MTO-Kraftstoff innerhalb EN228 E5 wiederfindet. Somit könnten auch Fahrzeuge, die noch nicht mit Partikelfilter ausgerüstet sind, deutlich sauberer betrieben werden, als dies heute der Fall ist. In diesem Zusammenhang erscheinen ebenfalls sehr sauber brennende, aber weit außerhalb der Norm liegende Kraftstoff-Exoten mit ggf. erheblichen Materialanforderungen wie beispielsweise Methylformiat (MeFo) oder Dimethylcarbonat (DMC) zwar als akademisch interessant, aber mit nur sehr geringen Umsetzungschancen für die automobile Praxis ausgestattet.

Weiterentwicklungen in der Normungsarbeit der EN228 sollten die hier gewonnenen Erkenntnisse zum Nutzen der Umwelt einbringen und eine deutliche Absenkung des Siedeendes ebenso vorsehen wie eine Begrenzung der höheren Aromaten (>C8) und ggf. einen Grenzwert für einen YSI.

Literatur

[1] Krämer, H., Borovsky, J., Bulc, A.: *Hochoktaniger paraffinischer Ottokraftstoff innerhalb EN228 E5.* In: Krahl, J., Munack, A., Eilts, P., Bünger, J. (eds.) Fuels Joint Research Group, Kraftstoffe für die Mobilität von Morgen, Band 30, pp.63-70. Cuvillier Verlag (2021).

Wasserstoff als alternativer Energieträger

Sebastian Rieß, Michael Wensing

Abstract

In the long term, it is a global goal to achieve a fully regenerative energy supply without fossil resources. Due to specific local differences with regard to the potential of renewable energies, a global energy economy is unavoidable in this context. From the perspective of energy efficiency, the direct use of electrical energy and thus the electrification of sectors and processes is the most economical way. Against the backdrop of a global energy scenario, however, this perspective must be part of a much more comprehensive economic analysis that also takes into account the aspect of energy distribution. Here, material solutions have a high potential. For short-distance energy transport, the use of existing power grids continues to be the solution. Over medium distances, however, hydrogen in pipeline networks is already gaining greater importance (European Hydrogen Backbone). For long-distance transport, liquid hydrogen derivatives such as methanol can be used in the existing fuel infrastructure.

A dedicated analysis including the power-to-power efficiencies of the different energy carriers (electricity, hydrogen, derivatives), their economic efficiency and availability resulting from the energy distribution, as well as the application-specific potentials and characteristics is needed to develop a strategy regarding which energy carriers can be used most sensibly in which sectors. Due to the local limitation of renewable energies and the range of potential applications of gaseous hydrogen in the sectors of industry, buildings and power generation, there seems to be a high potential of liquid derivatives in the transport sector. If such derivatives prove to be an economic solution anyway due to the energy distribution, a scenario arises in which the direct use of electrical energy as well as the use of gaseous hydrogen outside the transport sector seem to make more sense. In addition, there are advantages of liquid derivatives such as methanol with regard to drop-in potential, energy density and storage in mobile applications.

1. Einleitung

Die Welt sieht sich immer deutlicher mit Auswirkungen der globalen Erwärmung und des Klimawandels konfrontiert. Der Ausstoß von CO_2 ist längst kein rein technisches Thema mehr, sondern auch ein gesamtgesellschaftliches und politisches. Es stellt sich die globale Herausforderung, eine möglichst dekarbonisierte Energiewirtschaft einschließlich des Verkehrs- und Transportsektors zu schaffen. Eine grundlegende Voraussetzung und akute Herausforderung zur Erreichung dieses ambitionierten Ziels ist eine vollständig nachhaltige Erzeugung benötigter elektrischer Energie durch Solar-, Wind- und Wasserkraft. Aktuell zielt die Entwicklung im Verkehrssektor deutlich auf die Direktnutzung der elektrischen Energie in batterieelektrischen Fahrzeugen (BEV) ab. Es gibt jedoch Anwendungen, die entsprechend ihren Anforderungen bezüglich Energiemengen, Speicher- und Leistungsdichten nicht oder nur sehr schwierig elektrifiziert werden können. Hinzu kommen Fahrzeuge aus Bestandsflotten und industrielle Anwendungen, die ebenfalls nachhaltige Lösungen über die Elektrifizierung hinaus erfordern. Darüber hinaus wird die Energiedistribution auch über mitunter lange Verteilungswege eine zunehmende Rolle spielen. All diese Aspekte erlauben es nicht nur, sondern erfordern es, sich mit stofflichen Energiespeichern zu beschäftigen. Ein naheliegender Schritt ist hier Wasserstoff. Doch welche Rolle kann H_2 als alternativer Energieträger spielen?

2. Die Rolle von Wasserstoff in der Energiewende

2.1 Warum Wasserstoff?

Die Energiewirtschaft der Zukunft muss möglichst vollständig auf dekarbonisierten Energiequellen basieren. Die zentrale und prominenteste Rolle werden hierbei sogenannte regenerative Primärenergiequellen wie Sonne, Wind und Wasser einnehmen. Erzeugt wird daraus elektrische Energie. Nach einer Studie des VDE [1] kann eine Windkraftanlage mit einer Leistung von 3 MW und einer jährlichen Volllastlaufzeit von 2.000 Stunden 1.600 vollelektrische Fahrzeuge (BEV-Pkw) mit einer Jahresfahrleistung von 20.000 km versorgen. Wird die elektrische Energie genutzt, um mittels Elektrolyse Wasserstoff zu erzeugen, so können etwa 600 Brennstoffzellenfahrzeuge gleicher Jahresfahrleistung angetrieben werden. Bei Weiterkonversion des Wasserstoffs zu flüssigen E-Fuels können letztlich noch 250 Fahrzeuge versorgt werden. Aus dieser Perspektive betrachtet zeichnet sich ein eindeutiges Bild: Die höchste Energieeffizienz und den geringsten Primärenergiebedarf zeigt die direkte Nutzung elektrischer Energie in batterieelektrischen Fahrzeugen. Dabei kann ein BEV etwa 85 % der bereitgestellten elektrischen Energie für den Antrieb nutzen [2].

Eine unmittelbare Voraussetzung hierfür ist, dass die elektrische Energie dort zur Verfügung steht, wo sie zum Laden der BEV benötigt wird. Natürlich ergeben sich noch weitere Bedingungen und Einschränkungen wie beispielsweise die Verfügbarkeit von und der Zugang zu Ladesäulen/-stationen sowie Speichern für elektrische Energie. Es wird jedoch im Rahmen dieser Betrachtung davon ausgegangen, dass perspektivisch für die entsprechenden Antriebstechnologien die direkte Speicher-, Tank- und Beladungsinfrastruktur geschaffen wird und in ausreichendem Maße verfügbar und zugänglich ist.

Die Wasserstoff-Roadmap des Zentrums Wasserstoff.Bayern [3] konstatiert ein geringes Verfügbarkeits-Potential erneuerbarer Energien am Standort Bayern und den Status des Bundeslandes als Energieimporteur in einem zukünftigen nachhaltigen Energieszenario. Doch selbst die Distribution elektrischer Energie auf nationaler Ebene (Beispiel Nord-Süd-Trasse) ist bereits eine beträchtliche Herausforderung. Zwar schafft die Hochspannungsgleichstromübertragung technologisch gute Voraussetzungen für die Distribution elektrischer Energie, doch nichtsdestotrotz sind ökologische, gesellschaftliche und technologische Gesichtspunkte zu diskutieren. Vor diesem Hintergrund zieht die bayrische Wasserstoff-Roadmap den Schluss, dass ein stofflicher Energietransport mit Wasserstoff in einem Pipelinenetz anzustreben ist [3].

Eine analoge Überlegung kann neben der Landes- und Bundesebene auch auf globaler Grundlage angestellt werden. Auf Basis des Global Solar Atlas [4] zeigt sich deutlich, dass das Potential von Photovoltaik in Deutschland und Europa vergleichsweise begrenzt ist (Bild 1). In Regionen Afrikas, Nord-, Mittel- und Südamerikas sowie Australien hingegen ist beispielsweise ein enormes Photovoltaik-Potential zu verzeichnen. Auch bezüglich des Windkraftpotentials (Global Wind Atlas [5], Bild 2) ist eine lokale Konzentration der erreichbaren Leistung zu sehen. Hier ist das Potential in Küstenregionen besonders hoch, so beispielsweise vor allem in Skandinavien, Grönland und im Südosten Kanadas.

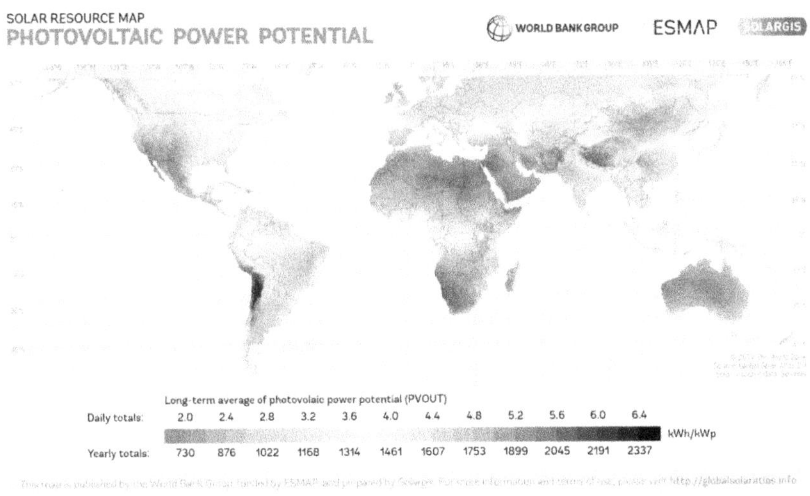

Bild 1: Standortabhängiges Photovoltaik-Potential [4]

Nicht nur die Wasserstoff-Roadmap Bayerns [3] sieht hier als zentralen Bestandteil der Energiedistribution aus Elektrolyse mit regenerativen Energien erzeugten Wasserstoff. Für die Verteilung des Wasserstoffs wird hier eine Anbindung an den „Europäischen

Wasserstoff-Backbone“ [6], ein von einem Konsortium von 20 europäischen Gastransport-Unternehmen vorgeschlagenes europäisches Pipeline-Netzwerk zur Wasserstoff-Distribution. Dabei können in Teilen bereits bestehende Gasleitungen genutzt werden. Wasserstoff hier als stofflichen Energieträger zu wählen ist naheliegend. Mit dem erwähnten Elektrolyse-Verfahren kann die Substanz direkt unter Nutzung elektrischer Energie aus Wasser hergestellt werden (Wirkungsgrade von 60 bis 75 % [2]). Wasserstoff hat einen hohen gravimetrischen Energiegehalt von etwa 120 MJ/kg und relativ vielfältige Einsatzmöglichkeiten. Nachteilig jedoch ist der sehr geringe volumetrische Energiegehalt des Gases von etwa 10,8 MJ/m³ bei Standardbedingungen.

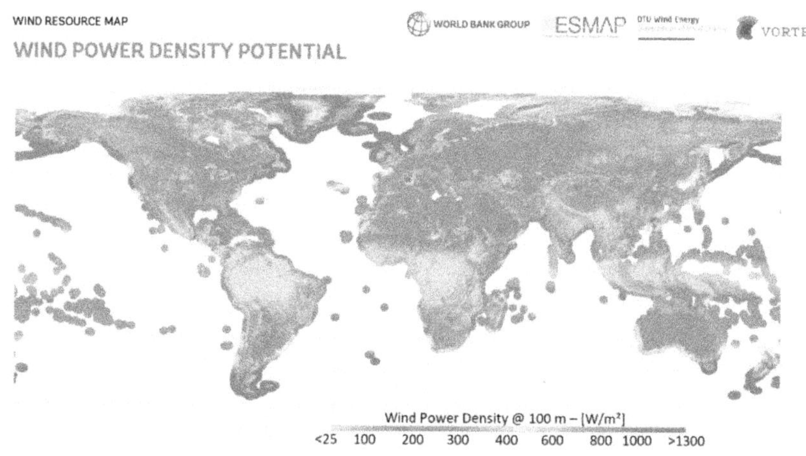

Bild 2: Standortabhängiges Windkraftpotential [5]

Wasserstoff kann darüber hinaus als Ausgangsbasis für die Weiterverarbeitung zu anderen stofflichen Energieträgern dienen. Hier sind vor allem Ammoniak und Methanol zu nennen, die direkt aus Wasserstoff hergestellt werden können. Ammoniak ist bei atmosphärischen Bedingungen gasförmig und zeichnet sich dadurch aus, dass es keinen Kohlenstoff enthält. Der spezifische Heizwert ist mit 18,7 MJ/kg deutlich geringer als der von H_2, der volumetrische Heizwert bei Standardbedingungen beträgt etwa 14,4 MJ/m³. Methanol enthält zwar Kohlenstoff, jedoch ist es möglich, den Alkohol unter Verwendung von Luft-CO_2 (direct air capture) direkt aus H_2 zu erzeugen, so dass er bilanziell CO_2-neutral sein kann. Der spezifische Heizwert des flüssigen Energieträgers liegt bei 21,1 MJ/kg, der volumetrische Heizwert bei Standardbedingungen bei etwa 16.800 MJ/m³.

Dort, wo elektrische Energie über lokale Stromnetze übertragen werden kann, ist Elektrifizierung die mit Abstand effizienteste und folglich auch wirtschaftlichste Lösung. Für Standorte wie Deutschland ist die lokale Verfügbarkeit elektrischer Energie aus regenerativen Quellen jedoch stark begrenzt. In einem kontinentalen und globalen

Energieszenario muss die Betrachtung der Wirtschaftlichkeit sowie Ökologie deshalb erweitert werden um den gewichtigen Bereich der Energiedistribution. Über Mittelstrecken (z.B. innerhalb Europas) stellt hier die Herstellung von Wasserstoff und die Verteilung über Pipeline-Netze eine Option dar. Für den globalen Langstreckentransport bieten sich darüber hinaus auf Wasserstoff basierende Flüssigderivate an, die in der vorhandenen Kraftstoff-Infrastruktur transportiert werden können und durch ihre hohe Dichte den wirtschaftlichen Transport großer Energiemengen ermöglichen.

In einer umfassenden Betrachtung müssen natürlich weitere Faktoren wie die Effizienz der letztendlich anzutreibenden Technologie, die Komplexität und Wirtschaftlichkeit (gerade mobiler) Tanktechnologien, Sicherheitsaspekte und End-of-Life-Szenarien neben den hier genauer betrachteten Aspekten berücksichtigt werden.

2.2 Was tun mit dem Wasserstoff?

Für die Betrachtung der Potentiale von Wasserstoff wird neben der direkten H_2-Verwendung ebenso die Verwendung wasserstoffbasierter Derivate herangezogen. Dabei beschränken sich die potentiellen Anwendungen nicht nur auf den Verkehrssektor, sondern erstrecken sich auch in die Bereiche „Industrie“, „Gebäude“ und „Stromerzeugung“.

Im Verkehrssektor hat Wasserstoff das Potential, direkt als Kraftstoff verwendet zu werden. Hierzu können Brennstoffzellen, Motoren, aber auch Turbinen eingesetzt werden. Brennstoffzellen haben tendenziell einen relativ hohen Wirkungsgrad, jedoch auch besondere Anforderungen an die Reinheit des Treibstoffgases. Brennstoffzellensysteme sind vergleichsweise komplex, ihre Belastbarkeit in Extremsituationen im Verkehr muss in der Breite noch erörtert werden. Im Bereich der Verbrennungsmotoren gibt es sowohl Konzepte der Saugrohreinblasung sowie der Niederdruckdirekteinblasung. Bei Saugrohreinblasung stellen sich Herausforderungen bezüglich Liefergrad, Aufladung und Sicherheitsaspekten aufgrund des Vorhandenseins von Brenngas im Ansaugtrakt. Bei der Niederdruckdirekteinblasung sind diesbezüglich Vorteile zu konstatieren, jedoch sinkt die effektive Tankausnutzung und damit die Reichweite im Vergleich zur Saugrohreinspritzung. Allen Konzepten der Verwendung gasförmigen Wasserstoffs ist ein hoher Aufwand zur Speicherung des Wasserstoffs gemein. So muss dieser entweder verdichtet (bis zu 700 bar) oder tiefkalt verflüssigt (etwa 20 K) werden, was einen erheblichen Energieaufwand bedeutet. Die Komplexität der entsprechenden Tanksysteme ist vergleichsweise hoch. Eine weitere Möglichkeit stellen im Verkehrssektor die schon erwähnten Wasserstoffderivate dar. Eine direkte motorische Verwendung von Methanol ist hier denkbar. Gerade im Schiffssektor kann Ammoniak in Dual-Fuel-Motoren eine Anwendung finden. Darüber hinaus können ausgehend von Methanol längerkettige Kohlenwasserstoffe als Drop-In-Kraftstoffe für die Bestandsflotte und Treibstoff für den Flugverkehr hergestellt werden. Prognosen zu Folge werden auch 2030 noch etwa 75 % der Fahrzeuge in Deutschland verbrennungsmotorisch angetrieben, so dass hier ein hoher Bedarf an derartigen Lösungen besteht [3]. Nachteilig ist die abnehmende Energieeffizienz der Herstellung im Bezug zur ursprünglichen elektrischen Energie. Wie bereits erwähnt muss diese in einer gesamtökonomischen Bilanz anderen Aspekten wie der Distribution entgegengestellt werden.

Im Industriesektor kann Wasserstoff sowohl als Rohstoff als auch bei der Bereitstellung von Prozesswärme eingesetzt werden. So besteht beispielsweise ein Bedarf in Hydrierprozessen und der Erzeugung einer reduzierenden Schutzatmosphäre [3]. Ein hohes Potential hat hier auch die Stahlindustrie, die in der Herstellung Wasserstoff zur Direktreduktion einsetzt und Erdgas substituieren kann [7]. Ebenso kann Wasserstoff in Raffinerieprozessen zur Chemikalienherstellung einen Einsatzbereich finden. Hier könnten durch den Einsatz grünen Wasserstoffs sogar Netto-Negativemissionen bezüglich des Kohlenstoffkreislaufs erzielt werden [7]. Vor allem bei Prozessen, in denen sehr hohe Temperaturen benötigt werden und eine Beheizung mittels Brenner sinnvoll oder unumgänglich ist, bietet sich das Potential des Einsatzes von Wasserstoff (oder synthetischem Methan als direktem Derivat) zur Bereitstellung von Prozesswärme. Konkrete Anwendungsfelder sind hier beispielsweise die Herstellung von Zement, Kalk, Glas, Pappe und Papier [3].

Im Gebäude-Sektor hängt das Potential bzw. der Bedarf des Einsatzes von Wasserstoff zur Erzeugung von Warmwasser und Raumwärme von der Sanierungsrate bei der lokalen Gebäude-Wärme-Peripherie ab. Wasserstoff selbst oder eines seiner Flüssigderivate können als Drop-In-Lösungen für Bestandsanlagen zur Heizung und Warmwasserbereitung fungieren [3]. Sowohl für den Industrie- als auch den Gebäudesektor kann es günstig sein, Wasserstoff über lokal vorhandene Gasnetze zu verteilen und Erdgas zu substituieren. Aufgrund des ähnlichen Wobbe-Index von Wasserstoff im Vergleich zu Erdgas ist dieses Potential gegeben.

Auch für die Stromerzeugung kann Wasserstoff künftig eine Rolle spielen. Aufgrund der Volatilität regenerativer Energiequellen kann die Herstellung von Wasserstoff aus Elektrolyse sowie ggf. die Weiterverarbeitung zu Derivaten als Energiespeicherlösung lokal genutzt werden. Die Speicherung des Gases oder seiner Derivate ist dabei in Tanks und/oder durch Einspeisung in bestehende Gasnetze verhältnismäßig einfach umsetzbar. Zur Rückverstromung können je nach Leistungsanforderung Brennstoffzellen oder Gaskraftwerke verwendet werden.

2.3 Einordnung der Energieträger im Verkehrssektor

Teils ambitionierte Szenarien stellen die Machbarkeit eines vollregenerativen Strom- und Wärmehaushalts in Deutschland perspektivisch in Aussicht [8]. Noch nicht berücksichtigt ist dabei der Energiebedarf des Verkehrssektors. Vor dem Hintergrund und der Herausforderung notwendiger Maßnahmen zur vollregenerativen Stromversorgung und angesichts der hohen Effizienz der direkten Nutzung elektrischer Energie scheint es sinnvoll, das limitierte Potential erneuerbarer Energien in großen Teilen in die Stromversorgung außerhalb des Verkehrssektors fließen zu lassen. Nichtsdestotrotz werden BEV vor allem bei Fahrzeugen geringen Gewichts und mit verhältnismäßig geringer durchschnittlicher Tagesfahrleistung (etwa unter 100 km) unbestreitbar einen signifikanten Anteil am zukünftigen Antriebsportfolio haben.

Vor dem Hintergrund der Notwendigkeit von Energieimporten kommt stofflichen Energieträgern zukünftig eine weiterhin bedeutende Rolle zu. Wasserstoff als unmittelbarster derartiger Energieträger hat dabei eine bedeutende Rolle als lokaler Speicher im volatilen regenerativen Energieangebot inklusive der Rückverstromung, in industriellen Prozessen (z.B. Stahl-, Zementindustrie) sowie als Drop-In-Lösung für Heizung und Warmwasserbereitung. Es muss bilanziell beurteilt werden, welche Mengen an Wasserstoff darüber hinaus und in Folge der Energiedistribution für den Verkehrssektor verfügbar bleiben. Eine ausreichende Verfügbarkeit vorausgesetzt, bieten sich dann Lösungen mit Brennstoffzellen und Verbrennungsmotoren für Anwendungen mit erhöhtem Fahrzeuggewicht und hoher durchschnittlicher Tagesfahrleistung an (z.B. Busse, Lkw, Offroad-Nkw).

Durch die erwartbare wirtschaftliche Verfügbarkeit von Wasserstoffderivaten wie Ammoniak und Methanol in einer globalen Energiewirtschaft und -distribution bietet sich auch angesichts der Potentiale und Notwendigkeit elektrischer Energie und gasförmigen Wasserstoffs außerhalb des Verkehrssektor weiterhin die Verwendung solcher Stoffe in Verbrennungsmotoren an. Gerade in Anwendungen mit besonders hohem Fahrzeuggewicht und hoher erforderlicher Reichweite (z.B. Schiff, Flugzeug) sind sie aufgrund der hohen Energiedichte und guten Speicherfähigkeit unumgänglich. Aufgrund dieser vorteilhaften Eigenschaften und abhängig von den in Folge der Energiedistribution vorhandenen Mengen bleiben sie dennoch auch weiterhin für Anwendungen in den zuvor benannten Bereichen leichterer Fahrzeuge mit geringeren Reichweiten eine Option hohen Potentials. Hinzu kommen das Drop-In-Potential sowie die Nutzbarkeit vorhandener Infrastruktur.

3. Zusammenfassung und Ausblick

Perspektivisch ist eine vollregenerative Energieversorgung ohne fossile Rohstoffe anzustreben. Aufgrund spezifischer Standortunterschiede hinsichtlich des Potentials erneuerbarer Energien ist eine globale Energiewirtschaft in diesem Zuge unumgänglich. Aus der Perspektive der Energieeffizienz ist die direkte Nutzung elektrischer Energie und damit eine Elektrisierung von Sektoren und Prozessen der wirtschaftlichste Weg. Vor dem Hintergrund eines globalen Energieszenarios muss diese Perspektive jedoch eingeordnet werden in eine wesentlich umfassendere Wirtschaftlichkeitsbetrachtung, die vor allem auch den Aspekt der Energiedistribution berücksichtigt. Hier haben stoffliche Lösungen ein hohes Potential. Für den Kurzstrecken-Energietransport bietet sich weiterhin die Nutzung vorhandener Stromleitungsnetze an. Über mittlere Distanzen erlangt jedoch bereits Wasserstoff in Pipeline-Netzen eine große Bedeutung (European Hydrogen Backbone). Für den Langstreckentransport bieten sich flüssige Wasserstoffderivate wie Methanol in der bestehenden Kraftstoffinfrastruktur an.

Es bedarf einer dezidierten Analyse unter Einbeziehung der Power-to-Power-Wirkungsgrade der verschiedenen Energieträger (Elektrizität, Wasserstoff, Derivate), ihrer aus der Energiedistribution resultierenden Wirtschaftlichkeit und Verfügbarkeit sowie den anwendungsspezifischen Potentialen und Charakteristika zur Entwicklung einer Strategie, welche Energieträger in welchen Sektoren am sinnvollsten eingesetzt werden können. Aufgrund der lokalen Begrenzung hinsichtlich erneuerbarer Energien sowie der Bandbreite potentieller Anwendungsgebiete gasförmigen Wasserstoffs in den Sektoren Industrie, Gebäude und Stromerzeugung scheint ein hohes Potential flüssiger Derivate im Verkehrssektor gegeben. Erweisen sich solche Derivate aufgrund der Energiedistribution ohnehin als wirtschaftliche Lösung, ergeben sich Szenarien, in denen die direkte Nutzung elektrischer Energie sowie die Nutzung gasförmigen Wasserstoffs außerhalb des Verkehrssektors sinnvoller erscheinen. Hinzu kommen Vorteile von Flüssigderivaten wie Methanol bezüglich Drop-In-Potential, Energiedichte und Speicherung in mobilen Anwendungen.

Literatur

[1] Petri R., Jaeger B., Lahdo N., Kesic M. Heusser D., *Antriebsportfolio der Zukunft – Ein Meinungsführer/-innen-Report aus Politik und Wirtschaft.* VDE Verband der Elektrotechnik Elektronik Informationstechnik e. V.: Frankfurt am Main (2021).

[2] Kramer U., *Defossilizing the transportation sector – Options and requirements for Germany.* FVV Forschungsvereinigung Verbrennungskraftmaschinen e.V.: Frankfurt am Main (2018).

[3] Runge P., Dürr S., Pfaffenberger F., Grimm V., Wasserscheid P., *Wasserstoff-Roadmap Bayern – Perspektiven und Handlungsempfehlungen zum Hochlauf der bayerischen Wasserstoffwirtschaft.* Zentrum Wasserstoff.Bayern (H2.B): Nürnberg (2022).

[4] Global Solar Atlas: https://globalsolaratlas.info/. Abgerufen am: 20.04.2022

[5] Global Wind Atlas: https://globalwindatlas.info/. Abgerufen am: 20.04.2022

[6] Wang A., van der Leun K., Peters D., Buseman M., *European Hydrogen Backbone How a dedicated hydrogen infrastructure can be created.* Guidehouse: Utrecht, NL (2020).

[7] Deutsche Energie-Agentur GmbH (Hrsg.), *dena-Leitstudie Aufbruch Klimaneutralität.* Deutsche Energie-Agentur GmbH (dena): Berlin (2021).

[8] Henning H.-M., Palzer A., *100 % erneuerbare Energien für Strom und Wärme in Deutschland.* Fraunhofer-Institut für Solare Energiesysteme (ISE): Freiburg (2012).

MAN Future Driveline

Stefan Buhl

1. Hintergrund

MAN sieht im Elektroantrieb von Nutzfahrzeugen einen entscheidenden Faktor auf dem Weg in die emissionsfreie Zukunft. Aufgrund der gesetzlichen Vorgaben zur CO_2-Reduktion im Bereich der schweren Lkw und den damit verbundenen potentiellen Strafzahlungen bei Nichterfüllung sowie den Vorgaben aus der Clean Vehicle Directive, sind neben dem technologischen Fortschritt speziell für diesen Bereich flankierende Maßnahmen seitens der Politik notwendig.

2. Kernaussagen

Die EU-Kommission hat mit dem Green Deal den Rahmen für den Verkehrssektor gesteckt: Um die CO_2-Emissionen bis 2030 um 30 Prozent zu senken und bis 2050 Klimaneutralität zu erreichen, muss in den nächsten 20 Jahren die Zahl der mit fossilen Kraftstoffen betriebenen Lkw und Busse reduziert und durch Fahrzeuge mit neuen Antrieben, die erneuerbare Energien nutzen, ersetzt werden. Die MAN Truck & Bus beobachtet deshalb die Marktentwicklung sehr genau und untersucht, welche Antriebstechnik sich vor dem Hintergrund der Gesamtbetriebskosten (total cost of ownership; TCO) und der nötigen Infrastruktur wirtschaftlich durchsetzen und bei unseren Kunden zum Einsatz kommen könnte. Im Zentrum unserer Bemühungen steht dabei der batterieelektrische LKW (BEV).

Gerade für den **städtischen und regionalen Verkehr** sehen wir die Elektrifizierung von Lkw und Bussen als gesetzt an, da dies im Hinblick auf die TCO am erfolgversprechendsten erscheint und sich die nötige Ladeinfrastruktur relativ einfach aufbauen lässt. Für den **Fernverkehr** (Lkw und Bus) sind derzeit verschiedene Szenarien denkbar. Wir gehen heute davon aus, dass ein typischer schwerer E-Lkw in Europa bei den Gesamtkosten bereits 2025 vor einem konventionellen Diesel-Lkw liegen wird. Eine mögliche Ergänzung zum BEV könnten die Wasserstoff-basierten Antriebe, vorrangig mit Brennstoffzellen-Technologie, sein.

Entsprechend dieser Analysen liegt der Fokus in der Serienentwicklung auf dem BEV (dargestellt in Abbildung 1). Die Entwicklung von Brennstoffzellen-Antrieben verhält sich dazu komplementär, baut also auf den batterieelektrischen Antrieben auf und ergänzt diesen mit einem H_2-On-Board-Charging, wodurch die eigentliche Batterie deutlich kleiner ausfallen kann.

Abbildung 1: Batterieelektrischer Schwerlast-LKW von MAN

Wesentliche Erfolgsfaktoren für den Einsatz von BEV oder Brennstoffzellen im Fernverkehr sind die Verfügbarkeit und die Kosten der jeweiligen Energieträger, die Ladeinfrastrukturentwicklung („High Power Charging" und H_2-Tankstellen) und der technologische Fortschritt. Generell gilt: Ein Umstieg gelingt nur, wenn entweder die Kosten für die Nutzung niedriger sind als bei den anderen Antriebsarten – Lkw und Busse sind Investitionsgüter, mit denen Unternehmen Gewinne erwirtschaften müssen – oder gesetzliche Rahmenbedingungen, wie z.B. erweiterte Lieferfenster, bevorzugte Zufahrt, Anreize für einen Umstieg setzen.

Dies vorausgesetzt, gehen wir davon aus, dass

- 2025 die Hälfte unserer neuen Busse alternativ angetrieben sein wird,
- bis 2030 rund 60% aller neuen MAN-Verteiler-Lkw mit Zero-Emission-Antrieben ausgestattet sein werden und
- der Alternative Antrieb im Fernverkehr einen Anteil von 40% haben wird.

3. Zusammenfassung

MAN Truck & Bus fokussiert sich bei der Entwicklung auf den batterieelektrischen Antrieb (BEV). Im urbanen Raum sehen wir diese Technologie als besonders sinnvoll an, weil sie eindeutige Vorteile bei den Gesamtbetriebskosten bietet. Weiterhin gehen wir davon aus, dass die Batterie-Technologie sich in den nächsten Jahren so weiterentwickeln wird, dass auch Fernverkehrsanwendungen möglich sein werden.

Wasserstoffantriebe (Brennstoffzelle) sehen wir als eine Ergänzung zum BEV. Der Erfolg der H_2-Technologie hängt stark von der Verfügbarkeit von kostengünstigem erneuerbarem Wasserstoff ab.

Role of Hydrogen as Energy Carrier and Fuel in the Energy Transition

Karsten Wilbrand

Kurzfassung

Der Verkehrssektor trägt erheblich zu den globalen CO_2-Emissionen bei. Tatsächlich macht der Straßengüterverkehr 9 % der weltweiten verbrennungsbedingten Emissionen aus und wird in den kommenden Jahrzehnten voraussichtlich weiter zunehmen.

Batterieelektrische und Wasserstoffantriebe sind beides effektive Lösungen, um den Schwerlastverkehr zu dekarbonisieren. In einer aktuellen Studie von Shell und der Technischen Universität Hamburg wurden Effizienz und Kosten der Energieträger-Lieferkette verglichen. Folgende Energieträger-Antriebsstrang-Optionen wurden untersucht: Batterie-elektrisch, gasförmiger Wasserstoff und flüssiger Wasserstoff mit Brennstoffzelle und ICE als Antriebsstrang. Als Referenzfall wurde der Dieselantrieb verwendet.

Das Papier gibt einen ganzheitlichen Überblick über die Produktion und Verteilung des Energieträgers bis zur Tankstelle und den Einsatz im LKW.

Die Studie berücksichtigt auch Einschränkungen aus der Perspektive des Energiesystems, d.h. den Bedarf an saisonaler Energiespeicherung und molekularen Energieimporten nach Europa. Wenn Wasserstoff in Zukunft in großem Maßstab nach Europa importiert werden soll, ist es dann besser, batterieelektrische Lastkraftwagen zu betreiben, oder sollte Wasserstoff direkt in einem FCEV- oder ICE-Antriebsstrang des Lkw verwendet werden? Die Antwort lautet: Beide Technologien werden in Zukunft gebraucht: batterieelektrische Lkw und Wasserstoff-Lkw. Es hängt vom benötigten Anteil der Wasserstoffumwandlung in Strom und vom Anwendungsfall der Lkw-Anwendung ab.

1. Introduction

The 2015 Paris Agreement defined a bold ambition to limit global warming to below 2 °C above pre-industrial levels and pursue efforts to limit it to 1.5 °C – in part by pursuing net carbon neutrality by 2050. In response, many countries, industries and individual organizations set targets to limit their carbon emissions and began developing plans on how to reach them.

Action is happening at global, national, regional, sector, and local levels, and there are many positive signs, but more can be done to address climate change. The United

Nations Environment Programme notes that: "On current unconditional pledges, the world is heading for a 3.2 °C temperature rise."

The challenge is particularly pronounced in six harder-to-abate sectors that, according to the International Energy Agency (IEA) [1], currently account for around 32% of global CO_2 emissions. These sectors share common characteristics, such as long asset lifespans, high energy dependency and complexity of electrification. As a result, decarbonisation of these industries will be slower, more investment-intensive and more technically demanding than other sectors. As decarbonisation happens more rapidly elsewhere, pressure and focus on harder-to-abate sectors is expected to increase. The transport sector is one of those harder-to-abate sectors and within this sector road freight is very significant in terms of CO_2 emissions.

2. Energy system transformation perspectives

Energy transitions towards a net-zero-CO_2 Europe by 2050 requires a wide ranging rebuild of the mostly fossil into a renewable energy system. All sectors need to contribute. The transport sector cannot be considered on its own but needs to be seen in the wider picture of the energy system transformation.

A clear trend towards electrification is foreseen not only for the transport sector but as well for industry and buildings. A strong increase in renewable power demand will be the consequence, see figure 1. However, even with strong targets to increase renewable power generation, the question remains how to provide renewable base load to the fluctuating wind and solar power.

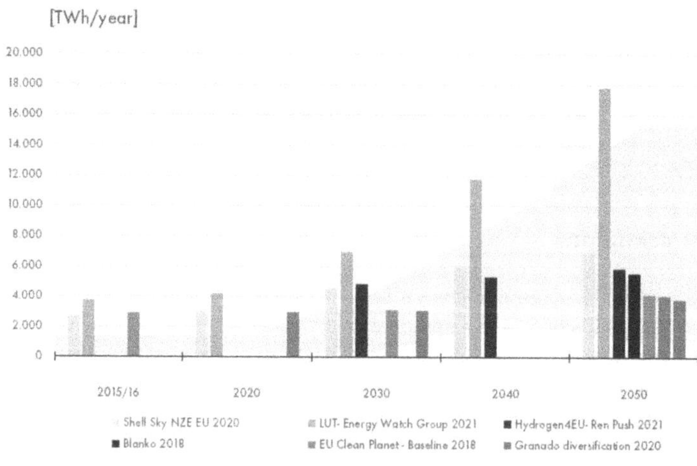

Fig. 1.: Total power demand in the EU – Scenarios based on external studies

For the new German government hydrogen is a key enabler of a successful energy transition [5]. Modern gas power plants should bridge the renewables' gap and these should be built "hydrogen-ready" in future. In addition, 10 GW electrolyzer capacity shall be available by 2030. At the end of 2021, the EU Commission has also published a package to decarbonize the gas grids using hydrogen. Thus, total hydrogen demand is expected to grow significantly in the EU to 2050, see figure 2.

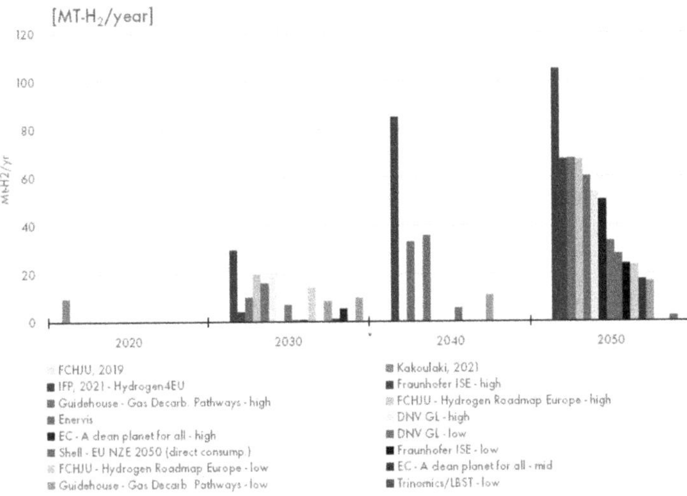

Fig. 2.: Total hydrogen demand in the EU – Scenarios based on external studies

Apart from local production of renewable hydrogen, it should be produced in regions where renewable power can be generated at low cost, sustainably, and where land is easily available. Regions as North Africa, Middle East, Patagonia or Australia would be preferred locations. Import to Europe can be realized depending on distance and need of final application via pipeline, as liquid H_2 by ship or as synthetic hydrogen derivatives or carriers like methanol, ammonia, or e-fuels.

Studies show that energy imports are expected to decline but could well still be between 20 % and 60 % in EU countries, see figure 3. Imports of renewable electricity and hydrogen and its derivatives are expected to increase significantly.

With growing demand for green molecules for energy-imports, storage and re-conversion to power, it seems sensible to consider a technology-open use of those molecules. Direct use in hydrogen trucks as well as indirect use via reconversion to electrons and use in battery electric trucks are the options to be considered in the following chapters.

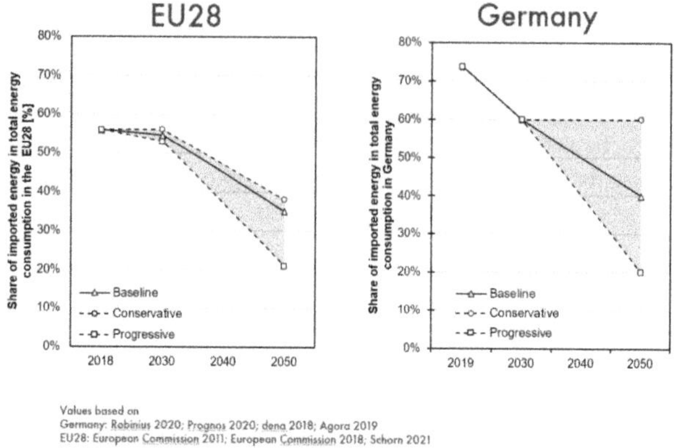

Fig. 3.: Future energy import dependency in EU28 and Germany [3]

3. Options to decarbonize heavy duty applications

There are three general pathways to decarbonize[1] powertrains, see figure 4. For decarbonization only renewable energy sources can be the starting point. Renewable electrons and renewable hydrogen can be used either directly with electric engines or in the case of hydrogen in a combustion engine.

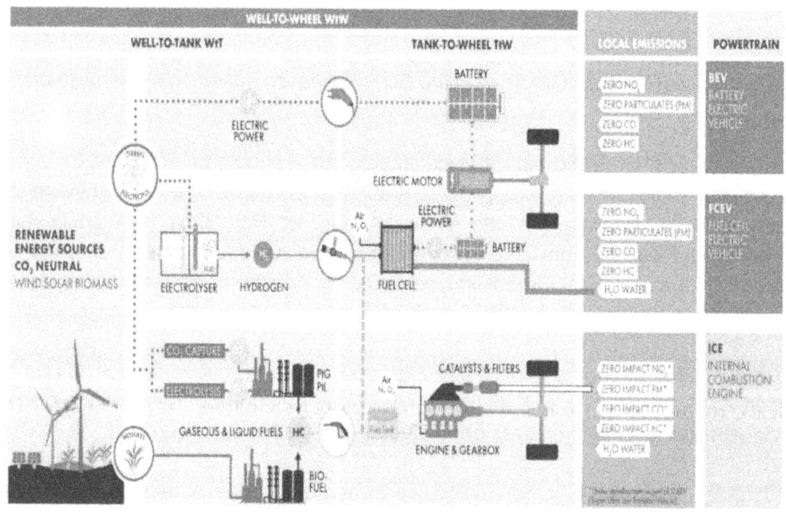

Fig. 4: Three pathways to ultimate clean vehicle drivetrains

[1] The term „decarbonized“ is used in the meaning of reducing fossil carbon. Sometimes the term “defossilized” describes the same in other literature.

Liquid or gaseous hydrocarbons can be generated from biomass (biofuels) or from renewable hydrogen and non-fossil CO_2 (e-fuels). These can be used in combustion engines, which need to be equipped with the ultimate clean emission aftertreatment system, which then would lead to "zero-impact" powertrains.

4. Energy carrier & powertrain study

Together with the Technical University of Hamburg ("TUHH") [3] and AVL [4] a model was developed to calculate the end-to end efficiency from solar/wind to the charger or the nozzle and from there to the wheel of a truck, see Fig. 5.

While TUHH calculated efficiency and cost of the energy carriers to the nozzle, AVL simulated battery electric and hydrogen powered trucks and calculated efficiency as well as cost of the different powertrains and vehicles.

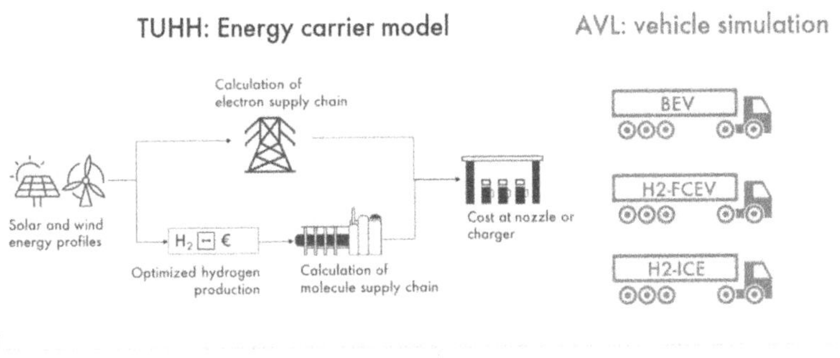

Fig. 5: Energy pathway model (simplified)

4.1 Energy efficiency calculations

The key results of the energy efficiency chain can be taken from Fig. 6.
The energy/hydrogen pathways in Fig. 6 (see superscripts in fig. 6) are described by the following

1: 50% direct supply by solar/wind energy + 50% reconversion of hydrogen into power (see 2)
2: 100 % reconversion of hydrogen (imported from Morocco by pipeline) into electricity by CHP gas-turbine
3: Gaseous hydrogen import from Morocco by pipeline
4: Liquid hydrogen import from Morocco by ship

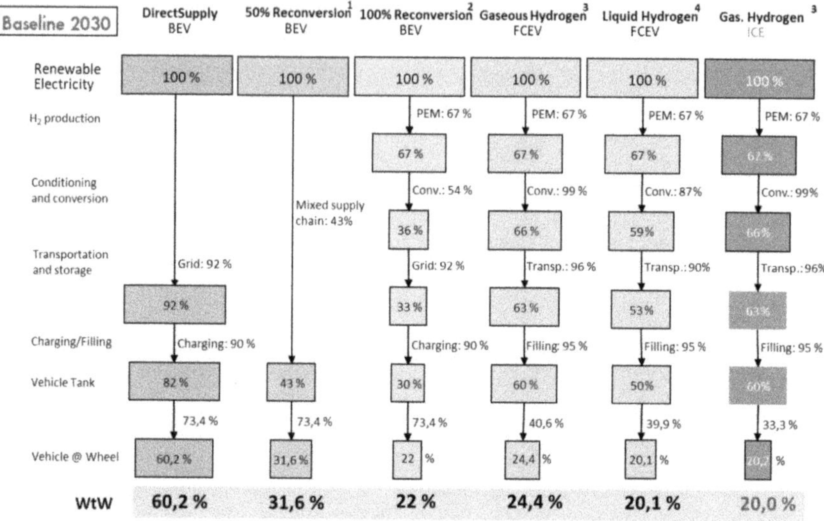

Fig. 6: Efficiency losses - comparison BEV and hydrogen

Naturally, due to the low number of conversion steps, the direct use of electrons in BEV ("Direct Supply") shows by far the highest WTW efficiency (60,2 %). However, renewable electricity in future will need a) seasonal storage, probably by hydrogen and b) imports via molecule, probably (also) hydrogen or hydrogen carriers (e.g., ammonia or LOHC). With growing demand for renewable energy, renewable electricity for battery electric trucks will also originate from reconversion of stored hydrogen. Fig. 6 assumes two cases: a) a case with "100 % reconversion" of imported hydrogen to electricity and a "50 % reconversion" case, which assumes that a BEV truck is charged with electrons of which half are stemming from renewable power generation (i.e., directly from local wind or solar generation) and by 50 % from hydrogen reconversion.

Reconversion of imported hydrogen into electricity causes substantial efficiency losses (WtW efficiency 22 % in the "100 % reconversion" case). The WTW efficiency of the hydrogen pathways ("gaseous hydrogen" and "liquid hydrogen" is approximately similar to the "100 % reconversion" case, with gaseous hydrogen showing a slightly better efficiency and liquid hydrogen showing a lower efficiency than the battery electric "100 % reconversion" case.

Use of gaseous hydrogen in a truck with a combustion engine shows the highest energy losses and hence a WtW efficiency of only 20 %.

4.2 Cost comparison of energy carrier supply options

Indicative cost of the energy carriers to the pump have been calculated as per Fig. 7.

Direct supply of electrons shows lowest cost. The "50 % reconversion" BEV cases show similar or higher energy cost at charger vs. the liquid hydrogen (LH_2) pathway. Compressed Gaseous hydrogen CGH_2 imported by pipeline is the cheapest hydrogen option. Cost of all hydrogen-based pathways come down significantly by 2030 vs. 2050 due to CAPEX reductions in solar PV, wind and electrolyzer technology.

The bars in Fig. 7 show the cost under progressive and under pessimistic assumptions. Uncertainties in the reconversion cases are higher due to a combination of uncertainties from electron and hydrogen pathways.

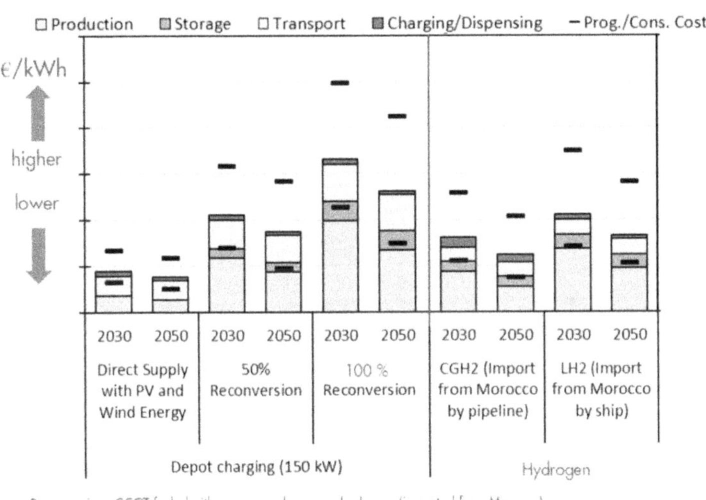

Fig. 7: Cost comparison of energy carrier supply options

References

1. IEA, CO_2 emissions by sector, World 1990–2018; CO_2 emissions from industry, transport and heavy-duty vehicles in the Sustainable Development Scenario 2000-2030
2. Decarbonizing Road Freight, GETTING INTO GEAR, Shell, Deloitte (2021)
3. TU Hamburg, Energy carrier supply chain model, Martin Kaltschmitt, Lucas Sens, Yannick Piguel (2021)
4. AVL, VECTO (Vehicle Energy Consumption Calculation Tool) cycle simulation (2021)
5. Mehr Fortschritt wagen. Bündnis für Freiheit, Gerechtigkeit und Nachhaltigkeit. Koalitionsvertrag 2021 – 2025 zwischen der Sozialdemokratischen Partei Deutschlands (SPD), BÜNDNIS 90 / DIE GRÜNEN und den Freien Demokraten (FDP), Berlin, den 24.11.2021

Solketal als Drop-in Komponente für nachhaltige Kraftstoffe

Julian Türck, Wolfgang Ruck, Jürgen Krahl

Abstract

The importance of fuel design for combustion engines will increase since it shows the potential to establish a low emission and Greenhouse Gas saving fuel. As a re-sult, the fuel will face more challenges. One important issue is the understanding and the research of interactions among fuel components. This article deals with a promising molecule, which is called Solketal. It can be produced from glycerin, which has a waste-based market. Considering the second educt acetone, Solketal can be a renewable biofuel. Due to its properties, it can be applicable in gasoline and diesel fuel.

1. Einleitung

Eine der großen Herausforderungen unserer Zeit ist die globale Erwärmung und das damit verbundene und erforderliche Ziel der Reduzierung der Emissionswerte. Wesentliche Faktoren sind dabei die Entwicklung und der Wandel der Mobilität. Aufgrund der Effizienz und Reichweite der Verbrennungsmotoren ist es nicht darstellbar, kurzfristig eine komplette Umstellung zu alternativen Konzepten zu vollziehen. Dem zufolge ist es für die Übergangsphase, oder auch für die Zukunft, wünschenswert, Kraftstoffe oder Motoren zu entwickeln, die kaum Emissionswerte generieren.
Kraftstoffe mit einem höheren CO_2-Einsparpotential sind Biokraftstoffe (z.B. Fettsäuremethylester (FAME) und Bioethanol). Um die Biokraftstoffquote zu erhöhen, ist es daher essenziell, die Wechselwirkungen der Kraftstoffe und der drop-in Komponenten untereinander zu analysieren. Beispielhaft für die Anforderung der Interaktion zweier bekannter Dieselbiokraftstoffe, dienen Oxymethylenether (OME) und die „Hydrotreated Vegetable Oils" (HVO). OME ist ein sauerstoffhaltiges Oligomer (Polyether), welches zur Verringerung der Rußbildung aufgrund der molekularen Sauerstoffkonzentration beitragen soll [1]. HVO hingegen sind Alkane, welche ausgehend von Pflanzenölen mittels Hydrierung hergestellt werden. Da der Anteil an unpolaren Kraftstoffen wie HVO zunehmen wird, werden polare Komponenten wie OME vor Löslichkeitsherausforderungen gestellt. Im Falle von OME ergibt sich eine Mischungslücke bei 30 bis 70% OME zu HVO [1]. Aufgrund der rußverringernden Eigenschaften von OME kann der dieselmotorische Prozess bei geringeren Emissionen betrieben werden, was eine Verringerung der dieselmotorischen Abgasnachbehandlung zur Folge haben kann [2].
Ein bedeutender Nebenstrom, der bei der Herstellung anfällt, ist Glycerin, welches das Grundgerüst der Öl- und Fettmoleküle bildet und aufgrund der drei Hydroxylgruppen eine verhältnismäßig größere molekulare Sauerstoffdichte aufweist. Glycerin selbst

kommt nicht als Kraftstoff in Frage (z.B. aufgrund der Acrolein-Emission), sodass mögliche glycerinbasierte Komponenten im ersten Schritt chemisch umgesetzt werden sollten. Vielversprechendere Ansätze der chemischen Derivatisierung von Glycerin werden von Bianchi et al. beschrieben [3]. Die Derivate sind attraktive Zündbeschleuniger, Klopffestigkeitsadditive, Viskosität- und Schmelzpunktverbesserer und Partikelemissionsverringerer [3]. Ein interessanter Kandidat, das Ketal Isopropylidenglycerin (Solketal), ist eine bereits bekannte Chemikalie, die als bipolares Lösungsmittel und auch als potenzielle Beimischkomponente in Benzin, Dieselkraftstoff und Biodiesel fungieren kann [4-6]. Der bisherige Einsatz von Solketal und dessen Derivaten liegt bis dato in der pharmazeutischen Chemie, wo es die Funktion als Synthesebaustein einnimmt [7]. Erste Untersuchungen zeigten, dass Solketal helfen kann, die Alterung von Biodiesel zu verlangsamen [8]. Dabei können polare drop-in Komponenten helfen, der durch die Alterung von Biodiesel und das damit verbundene Einbringen von Sauerstoff verstärkten Präzipitatbildung entgegenzuwirken, und letztendlich die Biokraftstoffquote zu erhöhen.

2. Solketal als grüner nachhaltiger Biokraftstoff

Solketal wird in die Gruppe der Acetonide und der Dioxolane eingestuft. Die Herstellung erfolgt mittels protonenkatalysierter Kondensationsreaktion von Glycerin und Aceton. In der Literatur sind eine Vielzahl verschiedener Synthesemöglichkeiten beschrieben (z.B. säurekatalysiert [9], Ionenaustauscher [10] etc.). Dabei entstehen zwei mögliche Konstitutionsisomere, welche in Abbildung 1 dargestellt sind.

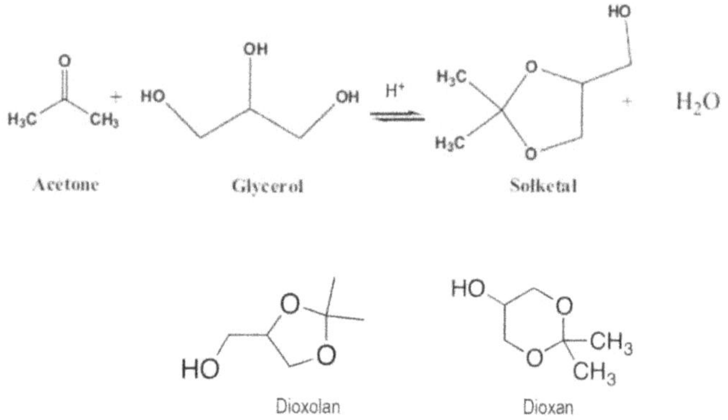

Abbildung 1: Reaktionsgleichung der Herstellung von Solketal (oben). Die zwei möglichen Konstitutionsisomere der Solketalsynthese: links: Dioxolan, rechts: Dioxan.

Der fünfgliedrige Ring, welcher zu ca. 97 % als Hauptprodukt entsteht, bildet die Dioxalanstruktur, wohingegen das sechsgliedrige Nebenprodukt eine dioxanbasierte

Struktur aufweist. Damit Solketal die Voraussetzungen für einen Biokraftstoff erfüllt, müssen alle Edukte der Herstellung biogenen Ursprungs sein. Glycerin gewährleistet dies aufgrund des pflanzlichen und tierischen Ursprungs. Ein biogenes Herstellungsverfahren für Aceton ist die Aceton-Butanol-Ethanol (ABE)-Fermentation, bei der aus Kohlenhydraten bakteriell Aceton, n-Butanol und Ethanol produziert wird [11]. Aufgrund der Klassifizierung als Biokraftstoff kann Solketal ein Kohlenstoffkreislauf zugeordnet werden (siehe Abbildung 2).

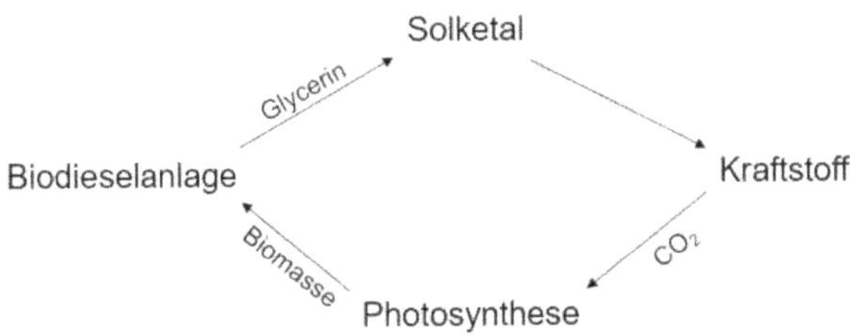

Abbildung 2: Kohlenstoffkreislauf eines Biokraftstoffes am Beispiel von Solketal.

Wenn Solketal als Kraftstoff eingesetzt wird, ergibt sich durch die motorische Verbrennung eine Produktion von CO_2. In der Photosynthese von Biomassen wird CO_2 als Ausgangsstoff benötigt. Wenn es sich bei den Biomassen um Rohstoffe für Biodiesel handelt (z.B. Raps, Soja, Sonnenblume etc.), wird aus dem gebundenen CO_2 ein Kraftstoff produziert. Damit sich der massenbilanzielle Kreislauf schließt, wird das anfallende Glycerin final zu Solketal umgesetzt. Solketal zeigt aufgrund der höheren molekularen Sauerstoffdichte ein polares Verhalten, verglichen mit den unpolaren fossilen Kraftstoffen. Somit sind vor allem in fossilen Dieselkraftstoffen bei höheren Solketal-Blendmengen Mischungslücken denkbar [6]. Ein Parameter, der diesen Sachverhalt thematisiert, ist die Permittivität (auch dielektrische Leitfähigkeit). Sie beschreibt die Polarisationsfähigkeit eines Materials durch elektrische Felder und kann somit mit der Polarität korreliert werden. Die folgende Tabelle 1 zeigt den Vergleich des Realteils der Permittivität $\varepsilon_r'(\nu)$ von fossilem Dieselkraftstoff (DK) mit Biodiesel und Solketal.

Tabelle 1: Realteil der Permittivität $\varepsilon_r'(\nu)$ von fossilem Dieselkraftstoff, Biodiesel und Solketal [6].

	fossiler DK	**Biodiesel**	**Solketal**
Realteil der Permittivität ε_r' (100 kHz)	**2,18**	**3,45**	**9,65**

3. Kraftstoffpotenzial von Solketal

Solketal ist nicht nur aufgrund seiner möglichen bio- und abfallbasierten Herkunft ein Molekül mit interessanten Kraftstoffeigenschaften. Dabei sind sowohl diesel- als auch ottomotorische Anwendungen denkbar und werden im Folgenden weiter beleuchtet.

3.1 Anwendung im Dieselkraftstoff

Das Ziel, polarere Komponenten mit einer höheren molekularen Sauerstoffdichte in den Dieselkraftstoff zu etablieren, kann im Wesentlichen auf folgende drei Potenziale heruntergebrochen werden.

3.1.1 Wirkung als polarer Anker für polare Alterungsprodukte

Um das CO_2-Einsparpotenzial mit einer Erhöhung der Kraftstoffquote von nachhaltigem Biodiesel gewährleisten zu können, ist es hilfreich, die Kraftstoffqualität des Biodiesels zu verbessern. Im Zuge der Biodieselalterung wird an den ungesättigten Bindungen Sauerstoff aufgrund der Autooxidation eingeführt [12]. Dieser Sauerstoffeintrag hat eine Oligomerisierung bei gleichzeitiger Erhöhung der Polarität zur Folge. Dabei können die polaren Alterungsprodukte ausfallen und ändern somit die chemischen und physikalischen Eigenschaften des Kraftstoffes [13]. Untersuchungen zeigten, dass polare drop-in Komponenten wie Solketal helfen können, dass diese Alterungsprodukte in Lösung bleiben [8]. Abbildung 3 zeigt die nachträgliche Zugabe von Solketal zu einem gealterten Dieselkraftstoff und dessen Aufklarung. Zudem zeigte der Dieselkraftstoff, welcher mit Solketal versetzt wurde und anschließend thermooxidativ im Rancimat gealtert wurde, dass eine Trübung ausblieb.

Abbildung 3: Der gleiche Dieselkraftstoff vor der Zugabe und nach der Zugabe von 3 Vol-% Solketal. Links: DK mit 3 Vol-% Solketal. Rechts: Trüber DK

3.1.2 Einfluss auf die motorische Verbrennung

Die Erhöhung der molekularen Sauerstoffdichte, und damit gleichbedeutend eine Veränderung der Verbrennung, kann mögliche positive Auswirkungen zeigen (siehe Einleitung). Prinzipiell zeigen oxidierte Komponenten wie Alkohole, Aldehyde und Carbonsäuren eine Absenkung der Cetanzahl. Dieser Sachverhalt wurde auch für Solketal bestätigt [6]. Erste emissionsanalytische Untersuchungen wurden mit Biodiesel-Solketal-Blends durchgeführt [14]. Durch Zugabe von Solketal erfolgte eine Abnahme der CO_2-, CO- und THC-Emission, wohingegen die NO_x-Emission zunahm. Dabei verringert Solketal die Viskosität, hilft aber dabei, die Kälteeigenschaften zu verbessern wie z.B. den „Cold Filter Plugging Point" CFPP [15].

3.1.3 Lösungsvermittlung für weitere polare Komponenten

Durch die Zunahme neuer potenzieller Kraftstoffkomponenten ergeben sich neue Wechselwirkungen und ein komplexeres Kraftstoffdesign. In Zukunft muss eine Balance zwischen Kohlenwasserstoffen (regenerativ oder fossil) und polaren sowie unpolaren Komponenten gefunden werden. Somit ergibt sich zwangsläufig, dass der Einsatz von amphiphilen Komponenten unerlässlich ist. Eine bekannte amphiphile Kraftstoffkomponente ist Biodiesel, der es erlaubt, polarere (zumeist oxidierte) Komponenten in Lösung zu halten [16]. Solketal zeigt auch, dass es lösungsvermittelnde Eigenschaften besitzt. So wird Vitamin C, welches nicht in Biodiesel löslich ist, durch Solketal in Lösung gebracht [4]. Solketal ist damit eine vielversprechende Kraftstoffkomponente.

3.2 Anwendung im Ottokraftstoff

Alkohole als Ottokraftstoffkomponenten sind marktetabliert und wurden ausgiebig untersucht. Alkohole wie z.B. Ethanol verbessern die ottomotorische Verbrennung und erhöhen vor allem die Klopffestigkeit (Oktanzahl). Prinzipiell zeigen Benzin-Solketal Blends ein ähnliches Verhalten wie Benzin-Ethanol Blends [17]. Alptekin et al. zeigten zudem, welchen Einfluss Solketal auf die Benzin-Norm EN 228 hat.

Auffällig ist vor allem, dass Solketal die sogenannte Gumbildung verringert [18]. Da Gum negative Auswirkungen auf die Motoreffizienz, Leistung, Emission und Haltbarkeit haben kann, kann eine Qualitätssteigerung durch Zugabe von Solketal erzielt werden. Zudem folgt die Gumbildung einem ähnlichen Mechanismus (radikalische Polymerisation) wie dem der Bildung polarer Alterungsprodukte im Dieselkraftstoff, sodass auch hier ein ähnlicher Einfluss von Solketal auf den Mechanismus möglich sein kann.

4. Zusammenfassung und Ausblick

Der Kraftstoff der Zukunft ist eine hochkomplexe Mischung mit verschiedensten Komponenten, der idealerweise optimale Kraftstoffeigenschaften zeigen wird. Solketal kann dabei eine wichtige Rolle spielen, da es zu einem CO_2-Einsparpotential beiträgt, das primär seiner Biogenität und im Speziellen der Kategorisierung als abfallbasierter Biokraftstoff geschuldet ist.

Insbesondere chemisch zeigt Solketal sehr interessante Kraftstoffwechselwirkungen und ist sowohl im Otto- als auch im Dieselkraftstoff einsetzbar. Solketal ist daher ein Forschungsfeld mit akademischem und anwendungsorientiertem Potenzial.

Tabelle 2: Einfluss eines Benzin-Solketal-Blends (10 Vol-% Solketal) auf die Benzin-Norm EN 228 [17].

	Prüfparameter	Benzin	Benzin-Solketal Blend	Einheit
	Heizwert	42,8	40,2	MJ/kg
	Dichte	743,8	773,8	kg/m^3
	Gumgehalt	0,4	0,2	mg/100 mL
	Oxidationsstabilität	>360	>360	min
	Siedeende	186,2	190,8	°C
	Destillationsrückstand	1	1	% (v/v)
	Distillation Percent % (E70)	41,1	37,6	% (v/v)
	Distillation Percent % (E100)	61,1	55,3	% (v/v)
	Distillation Percent % (E150)	92	84,2	% (v/v)
	Dampfdruck	63,1	57,7	kPa
	Sauerstoff	1,81	1,55	% (m/m)
Sauerstoffhaltige Verbindungen	Methanol	0	0	% (v/v)
	Ethanol	3,6	3,3	% (v/v)
	Isopropylalcohol	0	0	% (v/v)
	Iso-butylalcohol	0,7	0,6	% (v/v)
	Tert-butylalcohol	0	0	% (v/v)
	Ethers	2,3	2,1	% (v/v)
	Andere sauerstoffhaltige Verbindungen	0,3	0,2	% (v/v)
	Blei	<2,5	<2,5	mg/L
	Schwefel	7,5	6,8	mg/kg
	Benzol	0,75	0,68	% (v/v)
	Kohlenwasserstoffe (Olefine)	8	7	% (v/v)
	Kohlenwasserstoffe (Aromate)	34,9	34,3	% (v/v)
	Research-Oktanzahl (ROZ)	96,4	97	-
	Motor-Oktanzahl (MOZ)	85,8	86,2	-
	Kupferstreifen-Korrosionstest	1a	1a	Grad

Literatur

[1] J. Krahl, A. Munack, P. Eilts, and J. Bünger, *Handlungsfelder und Forschungsbedarf bei Biokraftstoffen*. Cuvillier Verlag, 2019.

[2] W. Leitner, J. Klankermayer, S. Pischinger, H. Pitsch, and K. Kohse-Höinghaus, "Synthese, motorische verbrennung, emissionen: chemische aspekte des kraftstoffdesigns," *Angewandte Chemie,* vol. 129, no. 20, pp. 5500-5544, 2017.

[3] C. Perego and D. Bianchi, "Biomass upgrading through acid–base catalysis," *Chemical Engineering Journal,* vol. 161, no. 3, pp. 314-322, 2010.

[4] F. Kerkel, D. Brock, D. Touraud, and W. Kunz, "Stabilisation of biofuels with hydrophilic, natural antioxidants solubilised by glycerol derivatives," *Fuel,* vol. 284, p. 119055, 2021.

[5] M. R. Nanda, Y. Zhang, Z. Yuan, W. Qin, H. S. Ghaziaskar, and C. C. Xu, "Catalytic conversion of glycerol for sustainable production of solketal as a fuel additive: A review," *Renewable and Sustainable Energy Reviews,* vol. 56, pp. 1022-1031, 2016.

[6] J. Türck *et al.*, "Solketal as a renewable fuel component in ternary blends with biodiesel and diesel fuel or HVO and the impact on physical and chemical properties," *Fuel,* vol. 310, p. 122463, 2022.

[7] L. Call, "Latente Funktionalität–eine Strategie zur Naturstoffsynthese," *Chemie in unserer Zeit,* vol. 12, no. 4, pp. 123-133, 1978.

[8] J. Türck, "Wechselwirkung und Einfluss von Solketal auf die Alterung von Fettsäuremethylestern," *Kraftstoffe für die Mobilität von morgen: 4. Tagung der Fuels Joint Research Group am 10. und 11. Juni 2021 in Dresden-Radebeul,* vol. 30, 2021.

[9] M. R. Nanda, Z. Yuan, W. Qin, H. S. Ghaziaskar, M.-A. Poirier, and C. C. Xu, "Thermodynamic and kinetic studies of a catalytic process to convert glycerol into solketal as an oxygenated fuel additive," *Fuel,* vol. 117, pp. 470-477, 2014.

[10] J. Esteban, M. Ladero, and F. García-Ochoa, "Kinetic modelling of the solventless synthesis of solketal with a sulphonic ion exchange resin," *Chemical Engineering Journal,* vol. 269, pp. 194-202, 2015.

[11] A. Kujawska, J. Kujawski, M. Bryjak, and W. Kujawski, "ABE fermentation products recovery methods—a review," *Renewable and Sustainable Energy Reviews,* vol. 48, pp. 648-661, 2015.

[12] F. Bär, M. Knorr, O. Schröder, H. Hopf, T. Garbe, and J. Krahl, "Rancimat vs. rapid small scale oxidation test (RSSOT) correlation analysis, based on a comprehensive study of literature," *Fuel,* vol. 291, p. 120160, 2021.

[13] A. Munack, L. Schmidt, O. Schröder, K. Schaper, C. Pabst, and J. Krahl, "Alcohols as a means to inhibit the formation of precipitates in blends of biodiesel and fossil diesel fuel," *Agricultural Engineering International: CIGR Journal,* 2015.

[14] E. Alptekin, "Emission, injection and combustion characteristics of biodiesel and oxygenated fuel blends in a common rail diesel engine," *Energy,* vol. 119, pp. 44-52, 2017.

[15] J. Delgado, "Procedure to obtain biodiesel fuel with improved properties al low temperatures," *EP Patent,* vol. 1331260, p. 45, 2002.

[16] S. Fernando and M. Hanna, "Development of a novel biofuel blend using ethanol–biodiesel– diesel microemulsions: EB-diesel," *Energy & Fuels,* vol. 18, no. 6, pp. 1695-1703, 2004.

[17] E. Alptekin, O. Ilgen, S. Yerlikaya, and M. Canakci, "Using Solketal-Gasoline Fuel Blends in a Vehicle with Spark Ignition Engine," *Chemical and Environmental Engineering,* vol. 7, no. 2, p. 93, 2016.

[18] C. J. Mota, C. X. da Silva, N. Rosenbach Jr, J. Costa, and F. da Silva, "Glycerin derivatives as fuel additives: the addition of glycerol/acetone ketal (solketal) in gasolines," *Energy & Fuels,* vol. 24, no. 4, pp. 2733-2736, 2010.

Nachhaltigkeit und Technologieakzeptanz – von der digitalen Filterblase der VUCA-Welt Wissenschaftsdialog mit der Gesellschaft[1]

Josef Löffl

Abstract

Traditional patterns of (scientific) communication are rapidly disintegrating in the age of social media. The frame story of the VUCA world requires a return to the power of connectable narratives. Scientific progress leads to the generation of new narratives that offer far-reaching opportunities for sensitising people to the demands of scientific thinking and acting. Of fundamental importance here is the transparent handling of complexity. In this way, the foundation is laid for a new kind of scientific dialogue.

Hinterher ist man immer schlauer – diese Binsenweisheit, so lapidar sie auch anmutet, fasst konzis unser Verhältnis zur Vergangenheit zusammen. Im Nachgang erscheint der Lauf der Dinge oftmals geordnet, zentrale Entscheidungen reihen sich einer Perlenkette gleich aneinander und alles scheint sich in eine größere Logik zu fügen.

Vielleicht trügt aber diese (scheinbare) Ordnung: Ist der angesprochene rote Faden unter Umständen nichts anderes als ein Zwangskorsett, das wir der Vergangenheit auferlegen, weil wir einen hohen Grad an Unsicherheit nicht ertragen können? Fakt ist jedenfalls, dass viele Definitionen, Termini und Bezeichnungen, die wir im Heute der Vergangenheit auferlegen, denjenigen, die diese Vergangenheit als ihre Zukunft durchlebten, fremd waren. Hätten wir eine Zeitkapsel zur Verfügung, die es uns erlauben würde, etwa in das Florenz des späten 14. Jahrhunderts zu reisen, würden wir sicherlich Befremden hervorrufen, wenn wir uns vor Ort nach dem Empfinden der stolzen Bewohner dieser Stadt als Zeitzeugen der Renaissance erkundigen würden. Entwicklungen können nach deren Durchlaufen wunderbar in Prozesse unterteilt und analysiert werden. Wie aber fühlt sich das Hier und Jetzt in einem Veränderungsprozess an? Ich vertrete die These, dass je tiefgreifender und umfassender sich eine Veränderung erweist, umso weniger nehmen wir diese wahr.

[1] Der vorliegende Beitrag findet sich als Erstabdruck in englischer Sprache im Tagungsband zum 12. Int. AVL Emissions and Energy Forum.

Es steht außer Frage, dass wir ein Zeitalter betreten haben, das sich grundlegend von den vorherigen unterscheidet. In jeder Epoche definiert sich Macht im Grunde als Kontrolle über einige wenige Schlüsselgüter – Getreide in Antike und Mittelalter, Kohle in der Zeit der Ersten Industriellen Revolution, Erdöl und fossile Brennstoffe fungierten als Kernressourcen der petrochemischen Revolution. In der Zeit der VUCA-Welt, die durch Volatilität, Unsicherheit, Komplexität und Ambiguität geprägt ist, fungieren Daten als Schlüsselelemente der Macht. Auf der Kontrolle dieser nicht-physischen Güter fußt ein globaler Informationskapitalismus, der klassische Mechanismen der Macht, die sich auf Arbeit, Boden, Geld gründen, ad absurdum führt.[2] Es bleibt aber nicht bei Veränderungen wirtschaftlicher Abläufe, wie wir sie auch in der Vergangenheit als Folge des technologischen Fortschritts gewohnt waren. Die Entwicklungen, die wir unter dem Schlagwort der Digitalisierung zusammenfassen, erweisen sich als radikalster Einschnitt in der Geschichte der Menschheit seit dem Übergang von der agilen Nomadenlebensweise als Jäger- und Sammler zur Sesshaftigkeit vor über zehntausend Jahren. Aus dörflichen Gemeinschaften erwuchsen Stadtstaaten, in denen Konzepte wie Politik, Wirtschaft, Religion und fast alles, was auch heute noch unsere Vorstellungswelt prägt, entstanden. Selbstverständlich handelt es sich bei dieser Darstellung um eine überspitzte Vereinfachung, da derartige Dinge keineswegs von Linearität und Gleichzeitigkeit gekennzeichnet waren. Kurzum: Die Welt war schon immer VUCA, nur ist die Dynamik heute eine andere, was zu teilweise paradoxen Situationen führt. Sinnbildliche Beispiele hierfür liefert unter anderem der österreich-amerikanische Historiker Walter Scheidel in seiner brillanten Studie über die Geschichte der Ungleichheit, in der u.a. thematisiert wird, dass im Jahr 2015 die reichsten 62 Personen auf diesem Planeten so viel besaßen wie die gesamte ärmere Hälfte der Menschheit.[3] Das alles ist möglich in einer Welt, in der de facto alle mit allen, bald alles mit allem und jeder mit allem vernetzt ist bzw. sein wird. Wiederum definiert sich dadurch auch Macht völlig neu: Wann zuvor war es jemals möglich, sprichwörtlich aus dem Kinderzimmer heraus Bewegungen mit globalem Impact zu starten? Gibt es Beispiele aus der Vergangenheit, die Analogien zu der Tatsache aufweisen, dass es heute durchaus möglich ist, mit einfachen technischen Möglichkeiten ohne großes Startkapital oder den Zugriff auf finanziell aufwendige Assets Unternehmen ins Leben zu rufen, die althergebrachte Geschäftsmodelle und die daran gebundenen Märkte scheinbar über Nacht disruptieren? Die Entwicklung unserer mentalen Antwort-/Lösungsmodelle auf die damit verbundenen Herausforderungen hinkt zwangsläufig hinterher, wie sich aus folgender Aussage von Annette Kehnel in ihrer Studie zur Geschichte der Nachhaltigkeit schlussfolgern lässt:

„Warum gehen uns trotz hektischer Suche nach Lösungen für die Herausforderungen des 21. Jahrhunderts allmählich die Ideen aus? Weil wir die Probleme der Zukunft mit Konzepten der Moderne lösen wollen. Moderne klingt zwar noch immer fortschrittlich und innovativ, doch ist diese Moderne historisch betrachtet mittlerweile mehr als zwei

[2] Vgl. Yvonne Hofstetter, Das Ende der Demokratie. Wie die künstliche Intelligenz die Politik übernimmt und uns entmündigt, 2. Aufl., München 2018, S. 408.

[3] Vgl. Walter Scheidel, Nach dem Krieg sind alle gleich. Eine Geschichte der Ungleichheit, Darmstadt 2018, S. 9-10.

Jahrhunderte alt. Das bedeutet, wir wollen die Herausforderungen des frühen 21. Jahrhunderts mit Konzepten lösen, die im späten 18. und 19. Jahrhundert entwickelt wurden."[4]

Das Empfinden einer Überforderung im Angesicht dieser neuen Gegebenheiten ist durchaus verständlich, da wir konzeptionell sprichwörtlich wie von gestern agieren. Das dadurch hervorgerufene Unbehagen steht im Mittelpunkt einer unlängst erschienen Studie von Armin Nassehi, in deren Rahmen er zu seiner Theorie einer überforderten Gesellschaft Folgendes ausführt:
„Die Digitaltechnik stellt eine besondere Herausforderung für den Latenzschutz dar, weil ihr Paradigmenwechsel darin besteht, direkt an den Latenzen der Gesellschaft anzusetzen. Die digitale Form der Informationsverarbeitung weiß mehr über uns als wir selbst – und das bereits, bevor es die elektronische Mediatisierung von Information in der heutigen Form gegeben hat. Das Bezugsproblem der Digitalisierung ist die Komplexität der Gesellschaft selbst. Das Unbehagen an der digitalen Kultur speist sich aus dem Sichtbarwerden dieser modernen Erfahrung. Es wird nun erst recht offensichtlich, dass die digitalen Möglichkeiten der flächendeckenden Beobachtung, die Rekombination von Daten und die Möglichkeiten des Kalkulierens die Akteure darauf stoßen, was sie zuvor latent halten konnten: Wie regelmäßig und berechenbar ihr Verhalten ist. Das Unbehagen an der digitalen Kultur ist womöglich der Spiegel, den diese Technologie der Gesellschaft vorhält: Sie überfordert die Gesellschaft und ihre Strukturen so sehr, dass sichtbar wird, was zuvor unter Latenzschutz stand."[5]

Die zuvor angesprochene Dynamik hebelt die (vermeintlich) schützende Latenz aus – in fast keinem Bereich wird dies so sichtbar wie im Sektor der Massenmedien, deren ursprüngliche Funktion in der Repräsentation spezifischer, z.B. politischer Milieus begründet lag: An ihre Stellen sind die sozialen Medien der Digitalen Revolution getreten. Diese, so Michael Seemann in seinen Ausführungen über die Macht digitaler Plattformen, repräsentieren nicht, sondern ermöglichen Verbindungen, was eine völlig andere Form der Politik begründet.[6] Die damit einhergehenden neuen Gesetzmäßigkeiten, die wiederum von Dynamik getragen werden – gefüttert vom unstillbaren Hunger nach Content auf Instagram, Facebook & Co. – sind nicht vereinbar mit grundsätzlichen Ansprüchen des Journalismus, wie Robert Simanowski betont:
„Die Begründung des Angebots an den Qualitätsjournalismus ist zugleich dessen Todesurteil. Wieviel Zeit darf ein Artikel zur intellektuellen Verarbeitung beanspruchen, wenn er sich keine drei Sekunden zum Erscheinen nehmen kann? Schnell geladen

---

[4] Annette Kehnel, Wir konnten auch anders. Eine kurze Geschichte der Nachhaltigkeit, 4. Aufl., München 2021, S. 11.
[5] Armin Nassehi, Unbehagen. Theorie der überforderten Gesellschaft, München 2021, S. 288.
[6] Vgl. Michael Seemann, Die Macht der Plattformen. Politik in Zeit der Internetgiganten, Berlin 2021, S. 368-369.

heißt auch schnell erledigt zwischen all den aufregenden Statusupdates. Die Faustregel der Aufmerksamkeitsökonomie lautet: Je einfacher man etwas liken kann, umso mehr Likes erhält es auch."[7]

Unter diesen Gegebenheiten rückt zwangsläufig Quantität an die Stelle von Qualität und formiert einen Rahmen, den wir als Filterblase bezeichnen dürfen, welcher von den Kernelementen Quantität, Dualismus und Tempo gekennzeichnet ist.[8] Das führt zu einem Mechanismus, der ebenso simpel wie wirkungsmächtig erscheint:
„Der numerische Populismus ist dem postfaktischen Emotionalismus verwandt: begründungslose Likes sind die technische Variation der gebetsmühlenhaften Wiederholung haltloser Slogans. So wie im realen Leben eine Lüge, die oft genug erzählt wird, für viele Wahrheit ist, so gewinnt eine Meldung auf Facebook dadurch an Gewicht, dass sie Gewicht hat: Man klickt immer auf die Angebote mit der höchsten Zahl und festigt so ihre Spitzenposition. Die Zahl ist ein Appell ans Gefühl, denn so viele können nicht irren, schon gar nicht, wenn meine besten Freunde darunter sind."[9]

Der Homo Digitalis hat nichts mit dem Irrglauben des Homo Oeconomicus zu tun: Ökonomen wie Robert J. Shiller haben mit ihren Überlegungen zum irrationalen Überschwang und anderen Thesen zu emotionsbasierten Entscheidungsprozessen den Weg für ein grundsätzliches Verständnis dieser Sachlage geebnet. Natürlich stellt sich in diesem Kontext die Frage, wie mit Aspekten dieser Art umgegangen werden kann. Meiner Auffassung nach ist zunächst der Forderung von Bernd Stegemann nach einer neuen Art von Aufklärung Genüge zu leisten, die die Gegebenheiten der Kommunikation in der Epoche transparent macht:
„Die Formenvielfalt, in der Öffentlichkeit entstehen kann, folgt der dialektischen Bewegung zwischen den Interessen der Mitteilung, den Interessen des Publikums und den technischen Medien, die für die Verbreitung verwendet werden. Die systemtheoretische Beschreibung leitet aus dieser Dreiecksbeziehung die Funktion von Öffentlichkeit ab. Diese besteht darin, Kommunikation sichtbar zu machen, so dass Kommunikation über Kommunikation entstehen kann. Damit unterscheidet sich öffentliche Kommunikation grundsätzlich von nicht-öffentlicher Kommunikation."[10]

Des Weiteren zeichnet sich die Gegebenheit immer deutlicher ab, dass die ursprüngliche Gatekeeping-Funktion eines Journalisten (oder eines Wissenschaftlers) im Grunde nicht mehr aufrecht zu erhalten ist. Auch die Herren über die digitalen Plattformen kämpfen damit, Regularien des Rechts auf ihren weltweit pulsierenden Kanälen adäquat umzusetzen, wie etwa den Ausführungen von Steven Levy am Beispiel von Facebook zu entnehmen ist:

---

[7] Robert Simanowski, Abfall. Das alternative ABC der neuen Medien, Berlin 2017, S. 104-105.
[8] Vgl. Robert Simanowski, Abfall. Das alternative ABC der neuen Medien, Berlin 2017, S. 25-26.
[9] Robert Simanowski, Abfall. Das alternative ABC der neuen Medien, Berlin 2017, S. 27.
[10] Bernd Stegemann, Die Öffentlichkeit und ihre Feinde, Stuttgart 2021, S. 59.

„Die Content-Standards zeugen von der Komplexität der Aufgabe. Weltweit gelten die gleichen Regeln, unabhängig davon, was in den jeweiligen Kulturen als zulässig erachtet wird. Die Standards gelten für alle Bereiche, die zu Facebook gehören, inklusive Newsfeed, Instagram, der Chronik-Funktion auf der Startseite und die Instant-Messaging-Dienste wie WhatsApp und Messenger. Die Regeln können sehr verwirrend wirken, geradezu jesuitische logische Höhenflüge erfordern. Manche sind einigermaßen unkompliziert. Da gibt es die Versuche, Stufen der Widerwärtigkeit zu definieren, etwa wenn es um die Bloßstellung menschlicher Eingeweide geht. Manches ist in Ordnung. Andere Varianten erfordern ein „interstitial", eine Warnung, die auf dem Bildschirm erscheint, ähnlich wie bei Sendungen im amerikanischen Fernsehen, bevor nackte Pobacken eingeblendet werden. Offen gezeigtes Durchbohren ist verboten. Ein gezeigtes Blutbad in den richtigen Ordner zu sortieren ist eine Ermessensentscheidung."[11]

Wie aber soll der Anspruch einer Qualitätssicherung in Einklang gebracht werden mit den aufgezeigten Gegebenheiten von Tempo und Quantität? Wir sollten das scheinbar Unmögliche nie kategorisch ausschließen, aber die Realisierung einer derartigen Zielsetzung muss als Quadratur des Kreises erscheinen. Dieser Eindruck verstärkt sich noch durch eine Gegebenheit, der wir das Attribut eines evolutionären Faktors zusprechen dürfen: Jeder von uns hat gerne recht und begrüßt eine Bestätigung seiner jeweiligen Meinung, ob dies nun im persönlichen Gespräch oder in einer der vielen digitalen Echokammern erfolgt. Tanit Koch führt dazu in einem aktuellen Sammelband über die Thematik der sog. Fake News Folgendes aus:
„Wir sind alle anfällig für Falschinformationen. Jeder von uns. Wer glaubt, er sei aufgrund eines höheren Bildungsgrades oder richtigen politischen Einstellung dagegen immun, ist gefährlich naiv. Alle Menschen neigen dazu, selektiv eher das wahrzunehmen und zu glauben, was sie in ihren bereits verfestigten Annahmen und Meinungen bestätigt, und wiederum das auszublenden oder abzutun, was ihre bestehende Auffassung hinterfragt. Diese uns innewohnende Tendenz nennt man „Confirmation Bias" und sie macht auch vor keiner Akademikertür halt. Wer also nur danach ruft, Kindern im Schulunterricht Medienkompetenz zu vermitteln, der übersieht den sehr viel größeren Teil der Bevölkerung, der die Schule längst hinter sich hat."[12]

Demnach ist es fahrlässig, einen hohen Bildungsstand per se als effizienten Schutzschild gegenüber Falschinformationen zu betrachten. An dieser Stelle lohnt es sich, in die Tiefe zu gehen: Es ist wichtig festzustellen, dass es sich bei uns von Anfang an um eine postfaktische Spezies handelt. Die Fiktion macht uns zu Menschen – auf diese Weise könnte die folgende These des israelischen Historikers Yuval Noah Harari zusammengefasst werden:

---

[11] Steven Levy, Facebook. Weltmacht am Abgrund. Der unzensierte Blick auf den Tech-Giganten, München 2020, S. 538.

[12] Tanit Koch, Customer Obsession. Ein Mittel gegen Desinformation?, S. 397, in: Tanja Köhler (Hrsg.), Fake News, Framing, Fact-Checking: Nachrichten im digitalen Zeitalter. Ein Handbuch, Bielefeld 2020, S. 383-399.

„Wir sind die einzigen Säugetiere, die mit zahlreichen Fremden zusammenarbeiten können, weil nur wir fiktionale Geschichten erfinden, sie verbreiten und Millionen andere davon überzeugen können, an diese Geschichten zu glauben."[13]
Allzu häufig steht der Begriff des content im Mittelpunkt der Diskussion um soziale Medien. Wir sollten uns vor Augen führen, dass wir als Menschen Narrative generieren, uns für diese begeistern oder davon abgestoßen werden. Diese Geschichten prägen unser Denken und Handeln in einer Art und Weise, deren Dimension prägnant von Samira El Ouassil und Friedemann Karig zusammengefasst wird:
„Diese Narrative sind deshalb so mächtig, weil sie nicht nur das Außen, sondern auch unser Innen bestimmen. Und das viel mehr, als den meisten von uns bewusst ist. In der Erzählung durch andere entwickeln wir überhaupt erst so etwas wie einen Geist, eine Idee von Identität. Nahezu alles, was wir heute das „Ich" nennen, stellt sich uns selbst und den anderen als Summe mehr oder weniger stimmiger Erzählungen dar. Wir sind, wer wir auf der Bühne anderer Bewusstseine zu sein glauben, genauer: welche Rolle wir dort von uns erzählen dürfen. Wenn das Sein dem Bewusstsein folgt und das Bewusstsein von ebendiesen Geschichten auf gewisse Kausalitäten hintrainiert wurde, liegt der Schlüssel zu einem gerechteren Sein im Kern dieser Narrative. Beleuchten und ändern wir Form und Inhalt unseres Erzählens, so beleuchten und ändern wir die Welt – und was es heißt, in ihr Mensch zu sein."[14]

Diese Ausführungen wirken mit Blick auf die aktuellen Anforderungen an den Wissenschaftsdialog zunächst sehr abstrakt. Ihre Bedeutung zeichnet sich aber dann deutlich ab, wenn wir sie mit der These von James Bridle in Verbindung setzen, die die Entwicklung einer neuen Technologie – meines Erachtens ist dies als Platzhalter für den wissenschaftlicher Fortschritt zu betrachten – als eine Erzeugung von Metaphern betrachtet:
„Mit der Herstellung eines Werkzeugs erzeugen wir ein bestimmtes Verständnis der Welt, das, auf diese Weise Gestalt geworden, in der Lage ist, bestimmte Wirkungen in dieser Welt zu erzielen."[15]

Wer etwas Neues schafft, generiert dadurch auch ein neues Narrativ, selbst wenn dies nicht zugleich ins Auge sticht. Einige dieser Narrative überdauern die Zeit und überlagern bisweilen den eigentlichen Erkenntnisgewinn, was diesem jedoch keinen Abbruch tut. Ein Paradebeispiel dafür ist etwa das Narrativ zu Isaac Newton, dem herabfallenden Apfel und der Entdeckung des Prinzips der Schwerkraft. Narrative dieser Art sind deshalb wichtig, weil sie eine allgemeine Anschlussfähigkeit herbeiführen, die Resonanz in den Kanälen der sozialen Medien produziert. Stellen wir uns doch an dieser

---

[13] Yuval Noah Harari, 21 Lektionen für das 21. Jahrhundert. Aus dem Englischen von Andreas Wirthensohn,
München 2018, S. 310.
[14] Samira El Ouassil, Friedemann Karig, Erzählende Affen. Mythen, Lügen, Utopien. Wie Geschichten unser Leben bestimmen, 2. Aufl., Berlin 2021, S. 18.
[15] James Bridle, New Dark Age. Der Sieg der Technologie und das Ende der Zukunft. Aus dem Englischen von Andreas Wirthensohn, München 2019, S. 22.

Stelle vor, das Newtonsche Szenario hätte sich unseren Tagen zugetragen: Zweifelsohne wäre der Sachverhalt „Herabfallender Apfel trifft brillanten Wissenschaftler und veranlasst diesen zu einem Erkenntnisgewinn epochalen Ausmaßes" ein Internet-Meme ohnegleichen. Natürlich wirkt dieser Sachverhalt auf den ersten Blick nicht nur banal, sondern geradezu kindisch. Aber wie wollen wir Menschen in einer von medialen Informationen überfluteten Welt tatsächlich für den Erkenntnisgewinn in der Wissenschaft sensibilisieren, wenn wir uns nicht diejenigen Phänomene zu eigen machen, die sich unter diesen Gegebenheiten als praktikabel erweisen? Es liegt auf der Hand, dass damit die Frage verbunden ist, wie Narrative dieser Art zu gestalten sind: Kann es eine „DIN-Norm" für Narrative der Wissenschaft geben? Ich vertrete in diesem Zusammenhang die These, dass in jedes dieser Narrative der Inhalt einfließen sollte, dass es sich nur um einen kleinen Einblick in komplexe Sachverhalte handelt und dass das jeweilige Narrativ lediglich neugierig machen soll, sich mehr Wissen über diese komplexen Sachverhalte anzueignen. Dies ist meiner Auffassung nach von entscheidender Bedeutung, um sich klar von einem Rahmen abzuheben, den ich wie folgt beschreiben würde:

1. Auf die Herausforderungen einer volatilen, unsicheren, komplexen und mehrdeutigen Welt kann es keine einfachen Antworten geben.
2. Das Narrativ, dass sich bei einem realiter unbeschränkten Zugriff auf Informationen automatisch eine auf wissenschaftlichem Denken und Handeln fußende Wissenschaftsgesellschaft herausbildet, hat sich als Trugschluss erwiesen.
3. Entscheidend ist, was in den Kommunikationsnetzwerken trotz des „Grundrauschens" anschlussfähig ist.
4. Als anschlussfähig erweisen sich simplifizierende Antworten auf komplexe Probleme.
5. Die Simplifizierung fungiert als Fundament jeglicher Form der Ideologie.

Unsere Narrative dürfen natürlich vereinfachen, müssen aber die Transparenz gewährleisten, aus der klar hervorgeht, dass es sich um eine Simplifizierung handelt. Ansonsten ist die Gefahr groß, dass sich aus einer vereinfachten Darstellung Ansprüche auf der Ebene von Wahrheiten ableiten, die in keiner Weise mit dem vereinbar sind, was wir als wissenschaftliches Denken und Handeln bezeichnen. Die gesamtgesellschaftliche Bedeutung dieses Ansatzes machte unlängst Alexander Bogner deutlich:

„Das Schicksal der Demokratie hängt an der Verbreitung eines typisch wissenschaftlichen Habitus, für den Skepsis, Selbstkritik und ein ausgeprägtes Misstrauen gegenüber absoluten Wahrheitsansprüchen grundlegend sind."[16]

[16] Alexander Bogner, Die Epistemisierung des Politischen. Wie die Macht des Wissens die Demokratie gefährdet, Ditzingen 2021, S.53.

Literatur

[1] Bogner 2021: Alexander Bogner, Die Epistemisierung des Politischen. Wie die Macht des Wissens die Demokratie gefährdet, Ditzingen 2021.

[2] Bridle 2019: James Bridle, New Dark Age. Der Sieg der Technologie und das Ende der Zukunft, München 2019.

[3] El Ouassil – Karig 2021: Samira El Ouassil, Friedemann Karig, Erzählende Affen. Mythen, Lügen, Utopien. Wie Geschichten unser Leben bestimmen, 2. Aufl., Berlin 2021.

[4] Harari 2018: Yuval Noah Harari, 21 Lektionen für das 21. Jahrhundert, München 2018.

[5] Hofstetter 2018: Yvonne Hofstetter, Das Ende der Demokratie. Wie die künstliche Intelligenz die Politik übernimmt und uns entmündigt, 2.Aufl., München 2018.

[6] Kehnel 2021: Annette Kehnel, Wir konnten auch anders. Eine kurze Geschichte der Nachhaltigkeit, 4. Aufl., München 2021.

[7] Koch 2020: Tanit Koch, Customer Obsession. Ein Mittel gegen Desinformation?, in: Tanja Köhler [Hrsg.], Fake News, Framing, Fact-Checking: Nachrichten im digitalen Zeitalter. Ein Handbuch, Bielefeld 2020, S.383-399.

[8] Levy 2020: Steven Levy, Facebook. Weltmacht am Abgrund. Der unzensierte Blick auf den Tech-Giganten, München 2020.

[9] Nassehi 2021: Armin Nassehi, Unbehagen. Theorie der überforderten Gesellschaft, München 2021.

[10] Scheidel 2018: Walter Scheidel, Nach dem Krieg sind alle gleich. Eine Geschichte der Ungleichheit, Darmstadt 2018.

[11] Seemann 2021: Michael Seemann, Die Macht der Plattformen. Politik in Zeit der Internetgiganten, Berlin 2021.

[12] Simanowski 2017: Robert Simanowski, Abfall. Das alternative ABC der neuen Medien, Berlin 2017.

[13] Stegemann 2021: Bernd Stegemann, Die Öffentlichkeit und ihre Feinde, Stuttgart 2021.

Sustainable Fuels for Maritime Shipping

Johann Wloka, Petra Rektorik, Alexander Knafl

Kurzfassung

Die maritime Energiewende sowie die Notwendigkeit zur Einsparung von CO_2-Emissionen sind sowohl durch die nationale als auch internationale Gesetzgebung (EU fit for 55 / IMO) erkannt worden und werden umgesetzt. Um die in diesen Regularien geforderten CO_2-Emissionsreduktionen zu erreichen, ist eine systematische und konsequente Defossilisierung notwendig. In der maritimen Schifffahrt kann dies nur durch den Einsatz nachhaltiger Kraftstoffe gelingen. Die MAN Energy Solutions SE ist hierbei ein Vorreiter für den Einsatz solcher Kraftstoffe und bietet bereits jetzt schon Lösungen hierfür an. Dieser Beitrag erörtert den Einsatz dieser Kraftstoffe an 2-Takt- als auch an 4-Takt-Motoren und bietet einen Überblick über vorhandene und zukünftige Entwicklungen.

1. Introduction

There is no doubt anymore that energy production needs to decarbonize rapidly in order to achieve the 1.5 °C goal from the Paris Agreement. For the automotive industry, the pathway, especially in Germany, but also in other parts of the world, seems to consist of a mix of battery electric vehicles and alternative fuels. Such fuels, like Fischer Tropsch (FT)-Diesel or FT-Gasoline, can be produced in regions of the world, where green energies (like wind and solar power) are more available and predictable.

For maritime shipping it is hardly possible to use batteries only. They can play a role as intermediate support for the engine or can be used in near costal area vessels, but for ocean going vessels liquid or gaseous fuels need to be used. The power density of such fuels is, even when it is partly much lower than for Diesel, still significantly higher compared to the power density of electric batteries.

FT Diesel can be an alternative but will be too expensive and furthermore not available as the on-road and off-road sectors will be absorbing it. Therefore, intermediate products, which are widely used and globally available, are a possible alternative. Methanol and Ammonia are such possible liquid fuels. Hydrogen is one key ingredient for the sustainable production of every fuel, especially the two mentioned. Hydrogen can also play a major role in green power generation – also for the shipping industry but mostly for stationary power generation. Therefore, the focus for the maritime shipping industry lies on both mentioned liquid fuels.

2. Methanol as a fuel

Methanol is the simplest alcohol and was already used in WW2 as an additive in gasoline fuel in order to improve the knocking behavior. Methanol can be also used as Diesel fuel but needs to be ignited by an external source – like a pilot spray of diesel.

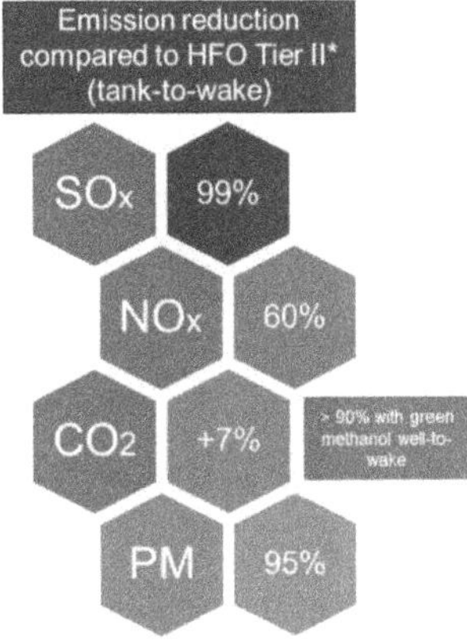

Fig. 1: Overview of Methanol Emissions

Methanol offers positive combustion performance. The emissions from methanol are all below the Diesel emission output. In case we use green methanol, the CO_2e [g/MJ] is 4. Using conventional production (from fossil methane gas), the CO_2 emissions are even worse compared to HFO (Heavy Fuel Oil). Highlights of the methanol combustion are the non-existing soot- and PM emissions. For the engine and the used technology (especially the injection system) multiple possibilities to inject the fuel to the combustion chamber are offered. There exist of course a few obstacles – the lower calorific value is approx. ½ of the value for Diesel Fuel, which makes bigger tank volume needed. There are other parameters of the fuel which are challenging (like the low flashpoint, the high toxicity, its corrosive potential – specially in combination with water, etc.). But there are technical solutions to overcome these obstacles.

The biggest challenge is more from the supply side. The yearly production capacity from E-methanol and bio-methanol from 2024-2025 onwards, where production capacity is already published, lies on 2.6 million tons of green methanol. That is enough to fuel approx. 150 Aframax Tankers for one year or 60 very large container vessels. As the CO_2 reduction potential is only valid for green methanol, see figure 1, the pick-up of production of green methanol needs to be accelerated rapidly. From the infrastructure point of view, everything is available what is needed in order to fulfill the needs of global bunker industry. Several bunker spots are already offering Methanol as a bunker fuel today, see figure 2.

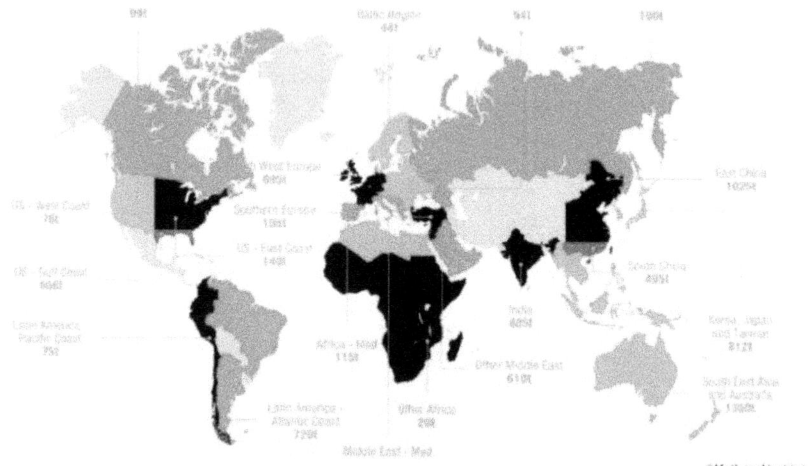

Fig. 2: Methanol on worldwide bunker spots, [1]

2.1 MAN solutions for Methanol

MAN offers a broad product portfolio for Methanol in order to support our customer's drive to decarbonize their business. The pioneer role takes MAN's 2 Stroke Engine division, which already now offers a range of several Methanol Engines ready for order, see figure 3.

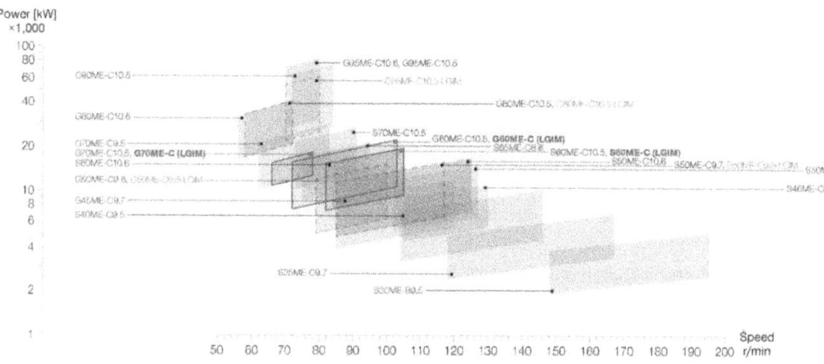

Fig. 3: MAN 2 Stroke Portfolio for Methanol

On the 4 Stroke side, retrofit packages are in discussion with several customers in order to retrofit existing diesel engines to Methanol operation. Specially the MAN medium speed portfolio ranging from cylinder diameters from 320 mm up to 510 mm are considered. Furthermore, MAN is participating on research projects for MAN 4 Stroke engines. There are for example investigations how to use Methanol on Marine Gensets for Methanol fueled container vessels in a project from the Danish Technology Institute with collaboration of MAN and the Danish Technology University, [2].

Besides the engine portfolio, MAN also supports the production of Methanol. In Haru Oni in Chile a demonstration plant for renewable Methanol & Gasoline Production is built. The CO_2 will be captured directly from the air. Hydrogen will be produced via a PEM electrolyzer. The Methanol synthesis will be done in an MAN reactor with a scope of 130.000 dm^3 / year. The renewable Methanol will then be further processed to green gasoline, see figure 4.

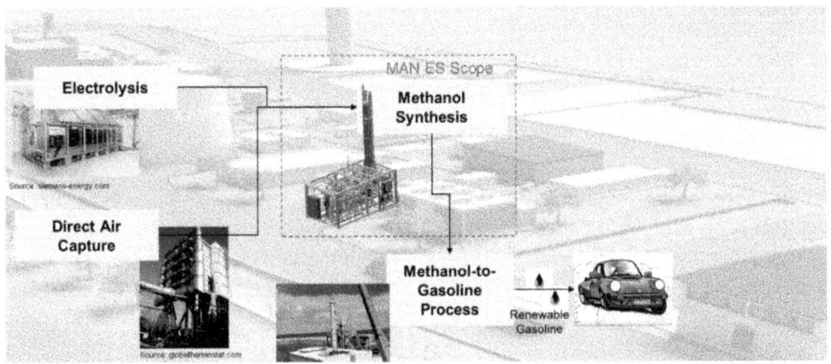

Fig. 4: Haru Oni Green Methanol production

3. Ammonia as a fuel

The biggest advantage of Ammonia as a fuel is the fact that it is carbon-free. With that a real decarbonization can take place. Several oil companies as ExxonMobil, Total etc. recognize Ammonia as a potential fuel for marine shipping. In contrast to Methanol, Ammonia is in gaseous phase under ambient conditions. In order to store Ammonia on a vessel the tank needs to be cooled down and pressurized to -33 °C and 6 bar. As the lower calorific value of Ammonia is lower compared to Methanol, also a bigger tank needs to be installed. Figure 5 compares the needed tank sizes to a standard diesel tank.

Comparable to Methanol, Ammonia can be used in a Diesel-Cycle. The high-pressure injection needs a pilot fuel to be ignited beforehand and burning. In this diesel flame the Ammonia is then injected and ignited. It burns then like in a diesel engine. As Ammonia is hard to ignite the pilot fuel amount is not comparable to state-of-the-art Dual Fuel Engines for LNG, which use lower than 1% of the energy to ignite the LNG. The pilot fuel oil demand is much higher. There are also other topics which are unknown for the Ammonia combustion, therefore MAN is participating in several research projects in order to gain the knowledge about the specifics of Ammonia combustion.

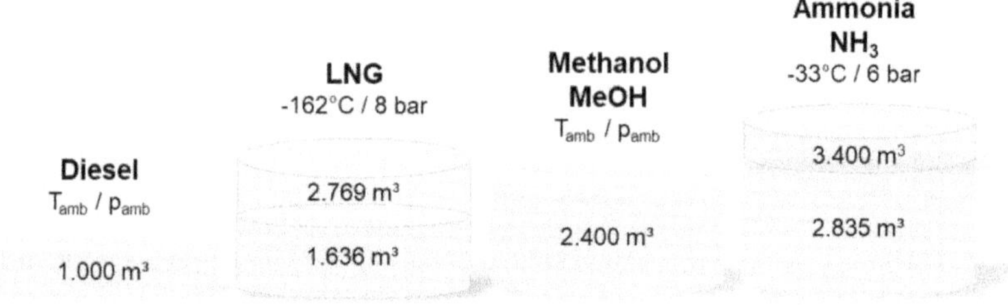

Fig. 5: Comparison of storage volume for the same energy amount

An overview on the Ammonia emissions can be found in figure 6. As with Methanol, practically no soot and SOx emissions are observed. On the other hand, as Ammonia carries a Nitrogen atom in the molecule, the NOx emissions will rise, but can be taken under control with established solutions for SCR, which MAN is also offering to its customers.

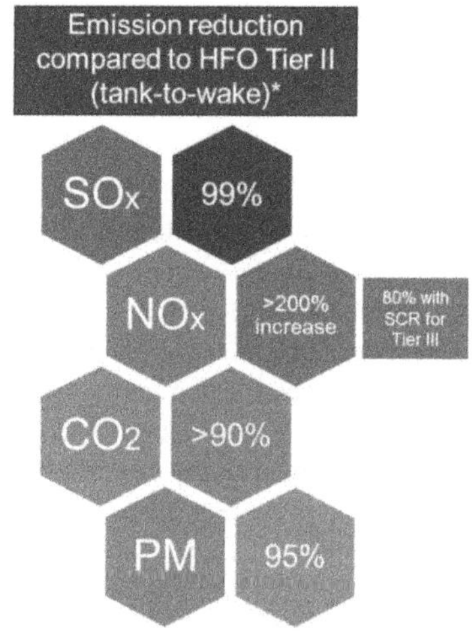

Fig. 6: Overview of Ammonia Emissions

3.1 MAN solutions for Ammonia

Both branches, the 2 stroke as well as the 4 stroke division of MAN are currently working on Ammonia engines. The test engines in the Copenhagen Research Centre are being reconfigured in order to investigate the combustion and all other systems (safety, fuel handling, etc.) on the engine with Ammonia. The solution is announced to be ready in 2024 and rolled consequently to the engine portfolio.

The 4 stroke division is investigating the ammonia combustion in a public funded project "AmmoniaMot" [3], with a broad industry consortium where the goal is to investigate the high pressure ammonia injection and combustion on a high-speed and medium-speed engine. Together with universities also fundamental investigations on the spray formation, ignition process and the combustion are performed.
The timeline for the introduction of NH_3 fueled engines is shown in figure 7.

Chronological order of market introduction

2-stroke-NH_3-engine on NH_3-Carrier

pilot-4-stroke NH_3-engine on NH_3-Carrier

4-stroke NH_3-engine on NH_3-Carrier

4-stroke NH_3-engine
other applications

Fig 7: Market introduction of MAN Ammonia fueled engines.

4. Conclusions

Two fuels are currently heavily discussed in the maritime industry, which have the potential to substitute Diesel and Heavy Fuel Oil (HFO) – Methanol and Ammonia. Both fuels are known chemicals that are nowadays widely used in different industrial branches. For both fuels the main advantage is their potential to reduce the CO_2 emissions drastically. On the other hand, such fuels also have setbacks that need to be considered. Toxicity, fuel storage and handling and, of course, safety requirements on board are a few to mention.

With its two engine divisions – 2 and 4 Stroke – MAN is offering different solutions for their customers to reduce their CO_2 footprint. Methanol 2 Stroke engines are already available, retrofit packages for 4 Stroke under development. Furthermore, MAN is investigating Ammonia as a fuel and developing solutions for its usage. With this MAN's strategy for decarbonizing the marine industry is developing and supporting the transition to sustainable maritime shipping and plays a significant role in stopping climate change, see also [4].

References

[1] Methanol Institute – Methanol on the Water, March 2022. Methanol Marine Fuel

[2] EUDP Project Methanol as fuel for marine diesel engines - Methanol as fuel for marine diesel engines | EUDP

[3] AmmoniaMot – Forschungsprojekt: AmmoniaMot – Wikipedia

[4] A Maritime Energy Transition – On the path towards climate neutral shipping. MAN Energy Solutions SE – White Paper download on www.man-es.com

Drop-in Compatibility Testing of 4-Stroke Marine Fuels: E-Fuels

Klaus Lucka, Simon Eiden, Chandra Kanth Kosuru

Abstract

With the introduction of different synfuels or e-fuels, the necessity to be drop-in capable has been an important factor. Being drop-in capable allows the product to achieve higher technology readiness level (TRL) and helps with the target of faster becoming climate neutral. Tec4Fuels has been part of several EU projects and German National projects in determining the drop-in capability of the novel fuels using the hardware in the loop technology and material compatibility testing. In the current topic, the adaptation of the hardware in the loop technology to test the novel-marine fuels will be discussed, and the results of baseline fuel testing will be presented.

1. Introduction

Hardware in the loop (HiL) principle is a core competence of Tec4Fuels in determining the compatibility of novel fuels and components from different sectors. In this testing all the fuel leading components of an automobile are connected in series. Finally, the injector is connected to a closed reactor in which the fuel is injected – collected and sent back to the tank. Thus, avoiding the combustion of the fuel allows the re-use of the fuel sample in a closed loop, see figure 1.

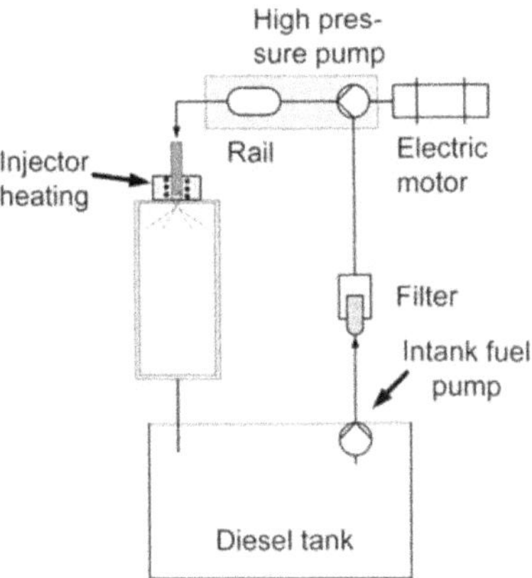

Figure1: P&ID diagram of a Hardware in the loop testbench

This re-usage of fuel produces additional stress on the fuel, leading to degradation. Thus, testing the components with the degraded fuel, and selecting a cyclic test run induces additional stress which can help in determining their endurance in a shorter period of time.

2. Adaptation of Testbench

For conducting the compatibility testing of marine fuels, the test methodology was adapted to concentrate on a single component which was found to be more critical. From the experience of several contacts of Tec4fuels, the critical issue was identified in the high-pressure pumps of the fuel injection system of a 4-stroke engine. Thus, considering the single component, the test bench was modified to easily induce different fuels and their blends and focus the testing on this aspect.

Figure 2 shows the adapted test set-up. The high-pressure pump in the picture is connected to a throttle which helps in creating the back pressure instead of an injector. Thus, the high-pressure pump can be operated in different conditions and the fuel is collected back to the tank after the throttle valve. Further the benchmarking tests are performed with the real-life marine fuel called the marine gas oil (MGO). The test conditions are fixed at 100 h of effective run time and usage of maximum of 30 liters of fuel.

Figure 2: Adapted test setup for the 4-stroke marine system

3. Results and Conclusion

The test was operated in a cycle, shown in picture 3, and the fuel quality was tested over time. This effect of high temperature and pressure has led to the degradation of the fuel quality. Although being a stable market available fuel, the test was successfully passed. Thus, the test bench is fully prepared for the usage of E-fuels. In an EU-funded project called IDEAL fuel the current setup will be used for testing the compatibility of Bio-HFO and its blends with the fossil MGO. This IDEALFUEL is a lignin based renewable fuel. Thus achieving shorter testing times and the usage of as less as 30 liters of the fuel sample helps in testing the development phased e-fuels efficiently and this testing will be used as a screening methodology for the future engine testing. Thus in the conference presentation, a short introduction on the project and the initial results of the testing will be presented.

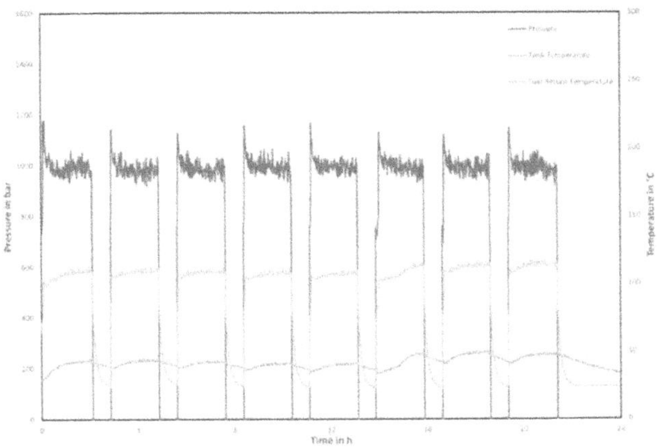

Figure 3: Test cycle of the HiL testing

Literature

[1] Hoffmann, H.; vom Schloß, H.: „Development of a Non-Engine Fuel Injector Deposit Test for Alternative Fuels (ENIAK-Project)" in Bartz, W.J.: Fuels: Conventional and Future Energy for Automobiles; 9th International Colloquium, January 15 - 17, 2013; TAE, Ostfildern, 2013, p. 613-615, ISBN 98-3-943563-04-7

[2] Hoffmann, Hajo: A Contribution to the Investigation of Internal Diesel Injector Deposits. Herzogenrath : Shaker Verlag GmbH, 2018. ISBN 978-3-8440-5953-3

[3] Hoffmann,Hajo; Sebastian Deist; Koch, Winfried; Lucka, Klaus: Development of a Non-Engine Fuel Injector Deposit Fuel Test: Results: TAE, Ostfildern 2015

This project has received funding from the European Union's Horizon 2020 research and innovation programme under grant agreement No 883753

Alternative Kraftstoffe und Einspritzsysteme

Enrico Bärow, Ingmar Berger, Michael Willmann

Abstract

In order to reach the decarbonization targets of the world economy, green hydrogen and derivative PtX fuels like Methanol and Ammonia will play a role in all sectors that cannot be decarbonized by direct electrification. In many of these applications, combustion engines can efficiently use these PtX fuels. The characteristics of the target applications influence the optimum combustion concept for the chosen PtX fuel. With the right set of engine technologies, a fuel switch during the lifetime of the engine is possible (fuel bridging). Woodward's adaptable injector family is the perfect platform to develop future proof combustion engine concepts for PtX fuels.

Kurzfassung

Um die Dekarbonisierungsziele der Weltwirtschaft zu erreichen, werden grüner Wasserstoff und abgeleitete PtX-Kraftstoffe wie Methanol und Ammoniak in allen Sektoren eine Rolle spielen, die nicht durch direkte Elektrifizierung dekarbonisiert werden können. In vielen dieser Anwendungen können Verbrennungsmotoren diese PtX-Kraftstoffe effizient nutzen. Die Randbedingungen der Zielanwendung definieren den infrage kommenden PtX-Kraftstoff. Mit den richtigen Motor-Technologien ist ein Kraftstoffwechsel während der Lebensdauer des Motors möglich („fuel bridging"). Woodwards Injektorfamilie ist die perfekte Basis für die Entwicklung zukunftssicherer Verbrennungsmotorenkonzepte für PtX-Kraftstoffe.

1. Introduction

The costs for the power generation from renewable energy sources have been plummeting for years. Now, they are cost competitive even without subsidies in most regions of the world. The increasing amount of wind and solar power requires new technologies to either store large amounts of electric energy or to convert it to chemical substances as synthetic fuels. It can be expected, that over the next years hydrogen technologies and Power-to-X (e.g., power to fuel) technologies will follow a strong cost decline as they scale up rapidly. On the basis of green hydrogen, it will be possible to create cost competitive synthetic fuels with zero carbon footprint. These synthetic fuels can be used to achieve a dramatic reduction of greenhouse gas emissions. This is especially the case in applications where a direct electrification and battery electric propulsion are not feasible.

The shipping industry, for example, is a large field of such applications. Assuming a "business as usual" scenario without major changes in propulsion technology, it is expected that the marine greenhouse gas emissions increase significantly up to 2050. However, to achieve the Paris agreement regarding the emission of greenhouse gases, the IMO aims to reduce emissions of the marine industry by 50% compared to 2008 in the same timeframe. Hence the pressure to decarbonize is increasing.
This publication gives an overview on how to choose the right Power-to-X (PtX) technologies for a given application and what the choices and benefits are, using a PtX fuel in an internal combustion engine.

2. Choice of PtX fuel

There are various PtX fuels in discussion. From the perspective of being GHG neutral they all have one commonality: they are produced from green hydrogen. Converting hydrogen to more complex chemicals improves the ease of storage but also increases the total costs and energy demand of production, especially if carbon capture is needed to generate hydrocarbon fuels. Fig. 1 shows, how much fuel volume is necessary to store a certain amount of energy. It can be seen, that for LNG, Methanol and Ammonia, the volume is almost two to three times as large as for Diesel or heavy fuel oil (HFO). For hydrogen, both liquid and compressed, the required volume is even larger. The increased efforts for the insulation of cryogenic or a pressure vessel for the compressed hydrogen worsens this comparison. The characteristics of the applications determine, if the benefit of lower fuel costs for hydrogen compensates the larger efforts for the storage.

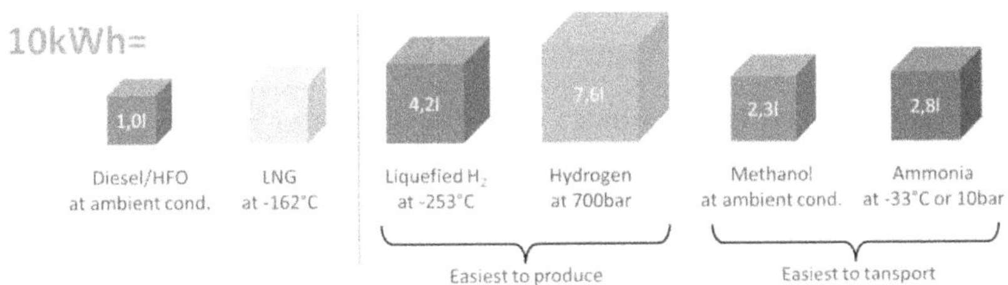

Figure 1: Required volume in litre of PtX fuel for storage of 10kWh (without storage tank)

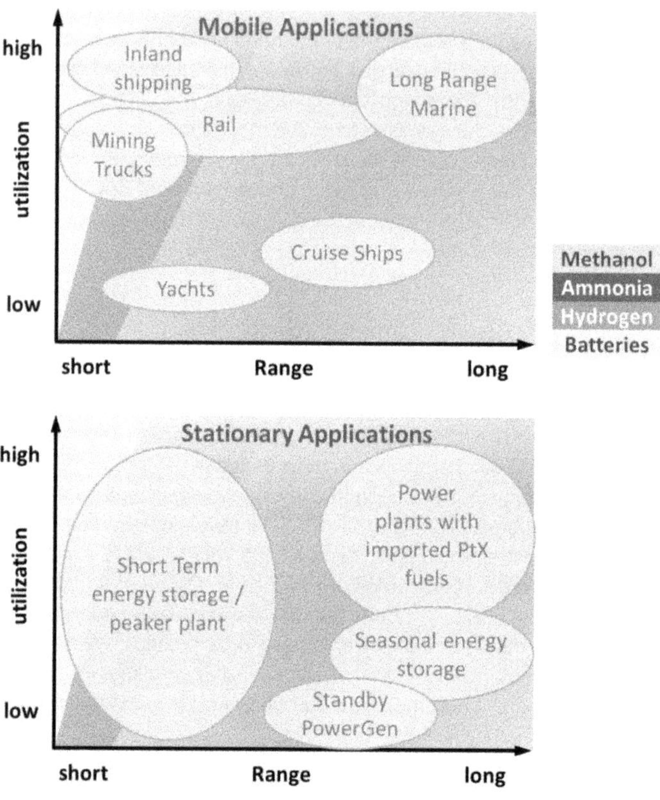

Figure 2: Schematic matrix for choice of power-to-X technology, top: for mobile applications, bottom: for stationary applications

This can be seen in Fig. 2. It shows a systematic approach to determine the best PtX technologies for different applications, divided in mobile applications (top) and stationary applications (bottom). For short range and especially highly utilized applications direct electrification or fuel cells might be the best solution. For many long-range applications a retrofitted or new internal combustion engine is more feasible. In these applications the utilization (fuel consumption) determines if the focus is on fuel cost or easier fuel handling. Ammonia produced with nitrogen is expected to be cheaper than Methanol produced with CO_2 [2]. Though, the current technological progress is further advanced for methanol combustion than for ammonia combustion. The use of methanol seems a reasonable first step and ammonia driven applications might become an attractive solution when the engine technology is proven.

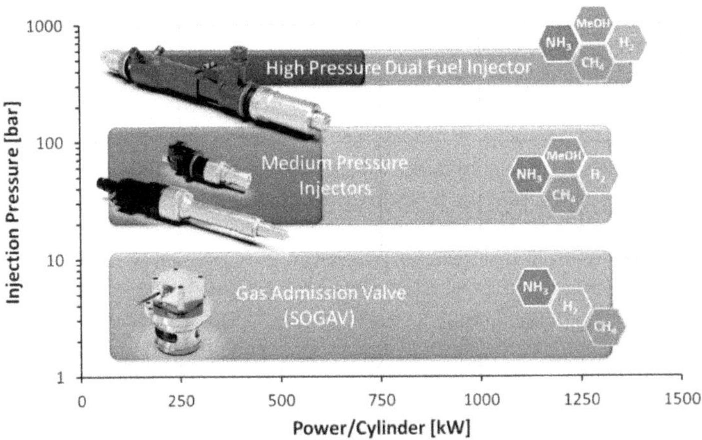

Figure 3: Overview of Woodward's PtX fuel injectors and valves

3. Fuel injection systems of Woodward

Woodward offers a broad range of injection technologies that are designed or adapted to inject synthetic fuels, either gaseous or liquid. Fig. 3 gives an overview of common injection techniques at different pressures. For many years the SOGAV port fuel injector (PFI) is used for applications where gaseous fuels need to be injected at low pressure upstream of the cylinder intake valves. The pressure range typically is from 5 to 10 bar. It offers a direct actuation and so it is an easy to use and robust injection valve. With a virtual sensor included, it is possible to monitor in real time the position of the valve position. This makes it a good solution for pure gas or dual fuel engines. Such low-pressure dual fuel engines rely on Woodward L'Orange's diesel injectors as ignition source. These can inject smallest quantities of fuel with extreme precision while also being able to run at full power in diesel mode.

For high pressure dual fuel applications, Woodward L'Orange recently introduced the new high-pressure dual fuel injector family [3, 4]. It allows the injection of both, diesel fuel at up to 2200 bar and a second (carbon neutral) fuel at 600 bar directly into the cylinder. This allows for a diesel like characteristic of the engine at highest efficiencies and power densities. The engine can be operated with both fuels at 100% load.

Woodward is continuously extending its portfolio of injection technologies to serve the broad range of PtX applications. To fill the gap between high pressure dual fuel and port fuel injection, the company is currently working on a flexible medium pressure injector platform for both, liquid as well as gaseous fuels. As can be seen in Figure 3, this injector can be designed to either inject the fuel at pressures of 20 bar or up to hundred bars.

As mentioned before, each application brings its own requirements regarding the fuel and the fuel injection. As diverse as the possibilities are, as diverse are the individual applications of PtX fuel injections.
Fig. 4 shows the current projects and inquiries at Woodward L'Orange. It can be seen that the main focus is on hydrogen, ammonia and methanol. For hydrogen, the injection at lower pressures is preferred – mostly as post fuel injection. Though, some projects also look at medium pressure direct injection in the range of 30-50 bar. For Ammonia the picture is more diverse. It ranges from gaseous port fuel injection at low pressures to injection at very high pressures as a liquid. When it is injected directly into the cylinder, it requires a strong ignition source. So, it is mainly combined with a diesel pilot in an HPDF injector. Other ignition techniques have a disadvantage here, because either they don't have enough energy to ignite the ammonia spray, or they are too big (like active prechambers). Whereas for hydrogen most engines are in the range below 300 kW per cylinder, Ammonia and Methanol seem to be good options even for larger engines up to the megawatt range. This might be due to the fact that large engines are often used in long range applications and thus seem to confirm the assumptions in the map Fig. 2. The main scope of many projects is to create a benchmark of different injection methods and combustion strategies to find the optimal solution for the specific applications.

During the coming decade the regulatory and market environment will undergo severe changes. Choosing a flexible engine technology today offers the chance to change to other fuels during the long lifetime of an engine. Different pathways for such a fuel bridging are shown in [5]. Woodward L'Orange injectors can be designed to allow a fuel switch with minor changes.

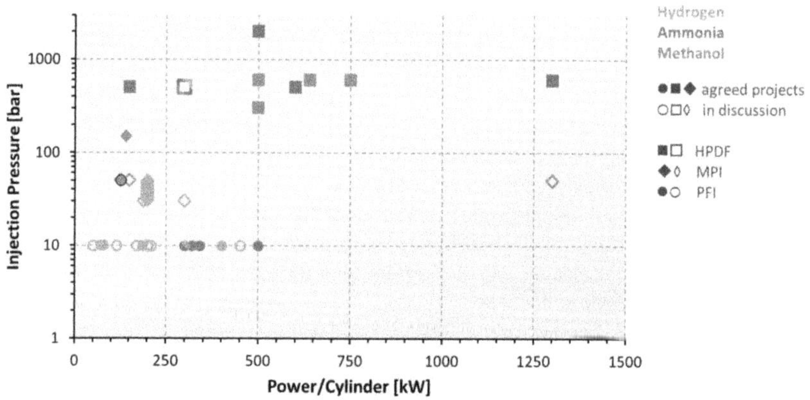

Figure 4: Snapshot of current projects and inquiries for injection systems at Woodward

References

[1] https://www.handelsblatt.com/unternehmen/erneuerbare-energien-gruener-wasserstoff-ist-zum-ersten-mal-guenstiger-als-wasserstoff-aus-erdgas/28251636.html, retrieved on 27th April 2022

[2] Zhao, Y., Setzler, B.,Wang, J., Nash, J., Wang, T., Xu, B., Yan, Y. (2019) "An Efficient Direct Ammonia Fuel Cell for Affordable Carbon-Neutral Transportation" Joule, doi: 10.1016/j.joule.2019.07.005

[3] Senghaas, C., Willmann, M., Berger I. (2019) "New injector family for high pressure gas and low caloric liquid fuels." 29th CIMAC Congress Vancouver

[4] Bärow, E., Willmann, M., Aßmus, K., Redtenbacher, C., Wimmer, A. (2019) "Operating Experience with a Combined High-Pressure Gas-Diesel Platform Injector". In H. Eichlseder (Ed.), 17. Tagung Der Arbeitsprozess des Verbrennungsmotors: 17th Symposium The Working Process of the Internal Combustion Engine (pp. 141 - 153). (IVT-Mitteilungen; Vol. 103). Verlag der Technischen Universität Graz

[5] Senghaas, C., Bärow, E. (2021) "Woodward L'Orange's New Injector Generation – An Adaptable Injector Family for Future Fuels", 9th AVL Large Engines TechDays, April 21 – 22, 2021

Visionen für zukünftige Luftfahrtantriebe

Friedrich Dinkelacker

Abstract

Future aviation should contribute to sustainability targets. Although current aviation concepts have already reduced emissions with respect to nitrogen oxides (NO_x), carbon monoxides (CO) and soot particulates, they still emit significant amounts here. Additionally, the aviation fuel produces carbon dioxides (CO_2) and water as main reaction products. CO_2 is seen as direct greenhouse gas. But, however, even the nitrogen oxide emission has most probably an additional impact in the same order on the radiation balance of the atmosphere, and soot particulates together with the water emission lead to contrail formation, which additionally has an impact on the radiation balance in the same order or even higher. Current environmental action is focused on the utilisation of sustainable aviation fuels (SAF) – for instance in the way, that aviation fuel might be produced with sustainably produced (green) electricity together with CO_2 from biomass or from carbon capture - so being one type of so called eFuel. This would reduce the direct CO_2 emission. However, the non-CO_2 contributions would remain. In this work, paths are investigated, where with advanced sustainable fuels also the non-CO_2 emission, especially the soot particulates and to some respect also the nitrogen oxide emissions will be addressed. This research vision is based on the hypothesis, that, with suitable advanced aviation eFuels, the ultra-clean Lean Prevaporized Premixed (LPP) combustion concept would be applicable, which has been proposed 30 years ago for aviation but has failed for kerosene. It would allow a dry ultra-low NO_x and particulate free combustion mode. First research steps are discussed to find good sustainably produced eFuels for the special utilization with the LPP combustion concept. The properties of such advanced eFuels need to be fully different from current kerosene substitute fuels. Next research questions are mentioned.

1. Situation von Luftfahrtantrieben und bisherige Alternativen

Auch wenn die Luftfahrt bisher erst 2-3 Prozent der weltweiten CO_2-Emissionen beiträgt [1], wird erwartet, dass dieser Anteil steigt, wenn beispielsweise die bodengebundenen Treibhausgas-Emissionen mit den Maßnahmen zur Erreichung des Pariser Klimaabkommens reduziert werden [2]. Eine Luftfahrtmobilität mit batterie-elektrischen Antrieben wird erforscht, ist aber mit derzeitigen Batterien (Energiespeicherdichte derzeit bestenfalls 300 Wh/kg = 1 MJ/kg) auf unter 400 km Reichweite begrenzt. Eine aktuelle Studie für einen Regionaljet erwartet eine Reichweite von knapp 1000 km, falls dann Batterien mit 700 Wh/kg (2.3 MJ/kg) verfügbar wären [3].[1] Mehr Hoffnung wird mit Wasserstoff und Antrieb über Brennstoffzellenwandler und E-Motoren oder Wasserstoff-Verbrennung [4] gesehen, wobei aber hier absehbar ein Drittel der Kabinenplätze aufgrund der großen Wasserstoff-Tanks benötigt wird. Für die Mittel- und Langstreckenflüge sind deshalb Konzepte basierend auf flüssigen Energieträgern mit ihren Energiedichten im Bereich von 40 - 45 MJ/kg so massiv im Vorteil, dass diese vermutlich auch in Zukunft dominieren werden. Akzeptabel erscheinen vielleicht alternative flüssige Energieträger wie Alkohole, bei denen die Energiedichten im Bereich von 30 MJ/kg liegen, was immerhin noch deutlich höher als bei Batterien ist.

Auch wenn bisherige Flugzeugantriebskonzepte im Vergleich zu früher deutlich sparsamer und sauberer sind, sind die Emissionen sowohl von CO_2 als auch der direkten Schadstoffe (besonders Stickoxide NO_x, Kohlenstoffmonoxid CO und Rußpartikel) und ihr Einfluss auf die Umwelt nicht zu vernachlässigen. Die gegenwärtige ökologische Diskussion fokussiert hier derzeit besonders auf die direkten Treibhausgasemissionen, also auf CO_2. Beste Langstreckenflugzeuge erreichen mit einem Verbrauch von 2,9 Liter Kerosin etwa 8 kg CO_2 pro 100 km (80 g CO_2 pro km) und Passagier.

Eine Vermeidung davon wäre mit sogenannten Sustainable Aviation Fuels (SAF) möglich, bei denen das Kerosin nicht fossil, sondern synthetisch hergestellt würde, wobei nachhaltig erzeugter Strom zur Wasserstoffsynthese und CO_2 aus Biomasse oder der Atmosphäre nötig wäre. Technisch ist dies möglich und erfordert nur relativ geringe Anpassungen am Tank oder Triebwerk der Flugzeuge. Nachhaltig für die nachhaltige synthetische Herstellung ist, dass die Herstellungseffizienz geringer ist als dies für Wasserstoff der Fall wäre.[2]

Allerdings lassen neuere Forschungsergebnisse der atmosphärischen Auswirkungen der Luftfahrt deutlich vermuten, dass auch die "Non-CO_2" Emissionen einen sehr starken überwiegend negativen Einfluss auf die Umwelt haben [5]. Ein Teil der Auswirkung betrifft die Emissionen aus Stickoxiden auf die Atmosphäre und hier insbesondere auch auf den Einfluss der Strahlungsbilanz zwischen eingestrahltem Sonnenlicht, das durch die Atmosphäre auf die Erde trifft und der Wärmeabstrahlung der Erde, die gegebenenfalls absorbiert wird, so dass diese Abstrahlung geringer wird, was zum Treibhausgaseffekt beiträgt. Auch wenn eine Quantifizierung dieser Einflüsse nicht so leicht ist und damit eine höhere Unsicherheit hat, gehen die Publikationen hier von einem

---

[1] Die Reisefluggeschwindigkeit ist hier mit Ma = 0.42 angenommen, bei den Mittelstreckenjets sind dagegen mindestens doppelt so hohe Geschwindigkeiten üblich und bei der hohen zeitlichen Auslastung der Flugzeuge auch nötig, die einen viel höheren Verbrauch oder eine entsprechend niedrigere Reichweite bedeuten würden.

[2] Auch der Preis ist für SAF-Kerosin bisher deutlich teurer als für fossiles Kerosin. Der Preis wird allerdings bisher für alle alternativen schadstoffarmen Antriebsarten noch wenig ernsthaft diskutiert und stellt für die Forschungsthematik bisher kein entscheidendes Kriterium dar. Vorteilhaft für SAF-Kerosin ist die etablierte Transportform als flüssiger Energieträger. Damit ist dies eine gute Handelsware, die weltweit am geeignetsten Ort produziert werden kann, die gut transportierbar und speicherbar ist, und die nahezu mit der vorhandenen Flugzeug- und Flughafeninfrastruktur auskommen würde, im Gegensatz beispielsweise zu Wasserstoff oder anderen Lösungen.

deutlich negativeren Einfluss der Emissionen auf die Umwelt aus, als es die direkten CO_2-Emissionen darstellen. Vereinfacht gesagt, wird von einem Einfluss der Stickoxidemissionen ausgegangen, der etwa gleich hoch wie der direkte CO_2 Einfluss des fossilen Kerosins ist (er sei im Folgenden mit 100 % Impact bezeichnet), während die Kombination aus Rußpartikeln und Wasseremission über die Kondensstreifenbildung und ihre Auswirkung zusätzlich 100 - 130 % Impact auf die Strahlungsbilanz bewirkt. Ein Ersatz des Kerosins durch synthetisch hergestelltes Kerosin würde (wenn Herstellungseinflüsse vernachlässigt werden) zwar zur Vermeidung der direkten CO_2-Auswirkung auf den Treibhauseffekt führen, weil der emittierte Anteil vorher bei der Herstellung der Biomasse oder beim Carbon-Capturing aus der Atmosphäre eingespart wurde,[3] aber die 200 - 230 % zusätzlicher schädlicher Klimaeinfluss durch NO_x und Kondensstreifenbildung bleiben, weil das Brennverfahren gleich bleibt.

Tabelle 1 zeigt hier eine eigene (vereinfachte) Abschätzung einiger für die Luftfahrt relevanter Parameter und des Einflusses auf die Luftfahrt für fossile Kerosin-Antriebe und für die bisher besprochenen Alternativen in grafischer Form. Das SAF-Kerosin hat hier zwar die großen Vorteile der Reichweite und der Tanktechnologie von bisherigen Flugantrieben, aber die Non-CO_2 Emissionen dominieren den Umwelteinfluss negativ.

Tabelle 1: Vergleich des bisherigen Luftfahrt-Antriebskonzeptes (Kerosin-Gasturbinen) als Basis mit einigen alternativen Luftfahrt-Antriebskonzepten (Batterie-elektrischem Fliegen, Wasserstoff-Brennstoffzellen-elektrisches Fliegen und nachhaltig produziertem Kerosin als eFuel mit Gasturbinen-Antrieb).

	Mass Range	Tank (Battery)	Energy-Efficiency Production	Direct GHG Emission	Non-CO2 Emission	Climate Impact (100%=CO2)	Challenges
Kerosene - GT (Basis)	+ +	+ +	fossil	- -	- -	300%	CO2, Emiss.
Battery - El	- -	-	+ +	+ +	+ +	30%	Mass too big
H2 - FC - El	+	-	+	+ +	+ +	50%	Tank, FC
SAF Keros. - GT	+ +	+ +	-	+	- -	200%	Emission

2. Die Vision der mager vorverdampften vorgemischten Verbrennung als Luftfahrtantrieb mit zukünftigen eFuels

Als Alternative zu konventionellen Flugantriebskonzepten schlagen wir die Mager-Vorverdampfungs-Vormischverbrennung vor (englisch: Lean Prevaporized Premixed LPP Combustion). In den 1990er Jahren wurden zu diesem Brennverfahren etliche auch

[3] Unter der Annahme, dass herstellungsbedingte Treibhausgasemissionen vernachlässigbar sind.

größere Versuchsreihen erforscht, bei denen das (damals fossile) Kerosin eingesetzt werden sollte [6, 7]. Dies gelang jedoch nicht befriedigend; was verständlich ist, weil Kerosin zu einem großen Teil aus langkettigen Kohlenwasserstoffen besteht, die sich bei den hohen Kompressionstemperaturen in einem Flugtriebwerk bereits nach sehr kurzer Zeit zersetzen und zur zu frühen Selbstzündung führen (z.B. [8]).

Beim LPP-Verfahren wird der zunächst flüssige Brennstoff bereits vor der eigentlichen Brennkammer in die komprimierte Luft zugemischt und hat etwas Zeit bis zur Brennkammer, sich mit der Luft zu mischen. Dies homogenisiert das Brennstoff-Luft-Gemisch und vermeidet inhomogene Bereiche, die für die Schadstoffbildung verantwortlich sind.

Ein LPP-Brennverfahren hätte die folgenden sehr großen Vorteile:

- Keinerlei Ruß-Partikelbildung, weil in der Brennkammer keine fetten Zonen mit Sauerstoffmangel auftreten, in denen in konventionellen Brennkammern die Rußpartikel entstehen.
- Nahezu keine Stickoxid-Bildung, weil durch die vorgemischte magere Verbrennung fast keine stöchiometrischen Flammenzonen auftreten, die für den Großteil typischer Stickoxid-Bildungsmechanismen in Flugtriebwerken relevant sind.
- Kleinere Brennkammern, weil die vorgemischte Verbrennung einen deutlich geringeren Brennraumbedarf hat als die bisherige konventionelle gestufte Verbrennung.

Obwohl auch die LPP-Verbrennung die Hauptprodukte (CO_2 und Wasser) erzeugt, ist die Umweltwirkung hier sehr viel geringer, weil keine Rußpartikel dabei sind, so dass die Kondensstreifenbildung ohne Kondensationskeime sehr viel geringer ist (diese wird etwa mit 30 % Impact im Vergleich zur reinen CO_2-Auswirkung abgeschätzt [5]). Zudem ist die NO_x-Emission aufgrund der fehlenden heißen Temperaturbereiche sehr viel geringer, bei stationären Gasturbinen oft unter 50 ppm, so dass auch dieser Impact-Anteil nahezu vernachlässigbar gering ist.

Wenn sowieso synthetische Ersatzkraftstoffe erzeugt werden sollen, die auf nachhaltiger Stromnutzung sowie bereits in der Biosphäre vorhandenem CO_2 basieren (sogenannte electro fuels oder eFuels), dann, so ist unsere Forschungsvision, können hier anstelle von Kerosin auch andere synthetische eFuels gesucht werden, bei denen die Eigenschaften für das sehr saubere LPP-Brennverfahren viel geeigneter als für Kerosin sind [9].

Im Bereich der großen stationären Gasturbinen, die insbesondere zur Stromerzeugung verwendet werden, ist der Übergang zur ähnlichen, dort als "Dry-LOW-NOx"-Verbrennung bezeichneten, fast schadstofffreien Verbrennung bereits in den 1990er Jahren vollzogen worden, weil das dort überwiegend verwendete Erdgas mit seinem sehr hohen Methan-Anteil sehr gut für eine Vormischverbrennung auch unter den entsprechend hohen Kompressionstemperaturen der Luft verwendet werden kann [10]. Für Flugtriebwerke ist aber das bei üblichen Bedingungen gasförmige Methan kein optimaler Kandidat, weil auch hier eine Tankproblematik besteht, und weil potentiell auftretende Reste mit unverbrannten Methan-Anteilen aufgrund des hohen Treibhausgaspotentiales von Methan für das Klima sehr nachteilig sind. Für andere synthetische Energieträger, idealerweise solche, die flüssig in Flugzeugtanks gespeichert werden können, gilt diese Einschränkung aber günstigenfalls nicht.

3. Eigene Forschung für LPP Verbrennung mit zukünftigen eFuels

Im Rahmen unserer Forschungsarbeiten im Excellence Cluster "Sustainable and Energy Efficient Aviation" (TU Braunschweig, Leibniz Universität Hannover, PTB Braunschweig und DLR) erforschen wir nun zusammen mit den Partnern (hier insbesondere die Gruppe von R. Fernandes und B. Shu an der PTB Braunschweig), ob es andere flüssige Energieträger gibt, die als synthetisch und nachhaltig herstellbares eFuel solche Eigenschaften haben, dass hier eine sichere Anwendung in einem LPP-Brennverfahren möglich wäre.

In unseren ersten Forschungsarbeiten haben wir hierzu ein zunächst etwas breiteres Screening potentieller synthetisch erzeugter flüssiger Kohlenwasserstoffe durchgeführt [9] und dann typische Zündverzugszeiten und typische Reaktionsraten (in Form laminarer Brenngeschwindigkeiten) untersucht. Erste Ergebnisse dieser mit der PTB Braunschweig gemeinsamen Arbeiten sind in [11, 12] veröffentlicht. Sie zeigen, dass beispielsweise kurzkettige Alkohole wie Propanol oder Butanol potentiell viel besser für LPP-Brennkammern geeignet sind als langkettige Kohlenwasserstoffe, wie sie im Kerosin enthalten sind. Dazu wurden als erster Schritt die Verbrennungseigenschaften genauer numerisch untersucht. Mit Hilfe eines neu entwickelten Reaktionsmechanismus für diese Gruppe wurden Flammengeschwindigkeiten und Zündverzugszeiten bei triebwerksrelevanten Bedingungen (860 K und 36 atm) berechnet und schließlich hinsichtlich ihrer Eignung als LPP-Brennstoff bewertet. Bild 1 zeigt berechnete laminare Flammengeschwindigkeiten, die als fundamentales Maß der effektiven Reaktivität des Brennstoff/Luftgemisches gesehen werden können. Während sie für den LPP-Betrieb beim bisherigen Flugkraftstoff viel zu hoch ist, sind die untersuchten Propanol- und Butanol-Isomere hier vielversprechend.

Ebenso wurden die Zündverzugszeiten berechnet, die für den LPP-Einsatz hinreichend groß sein müssen (mindestens größer als die Verweilzeit im Mischbereich eines LPP-Brenners, die im Bereich einiger ms liegt). Bild 2 zeigt, dass hier besonders das Iso-Propanol und das tert-Butanol gute Kandidaten für einen erfolgreichen Einsatz als LPP-Kraftstoff sein können, weil hier die Zündverzugszeiten hinreichend hoch sind, während die Daten für Jet A-1 viel zu kurze Zündverzugszeiten zeigen. In [12] sind hierzu auch experimentelle Daten zu Zündverzugszeiten dargestellt, die in der Hochdruck-Zündverzugskammer (High Pressure Shock Tube) der PTB Braunschweig im Bereich von 1100 bis 1500 K für 10 bar und teilweise auch bis 40 bar sehr genau vermessen werden konnten. Hierdurch ist eine Entwicklung und Anpassung der Reaktionsmechanismen für diese Bedingungen möglich geworden [12].

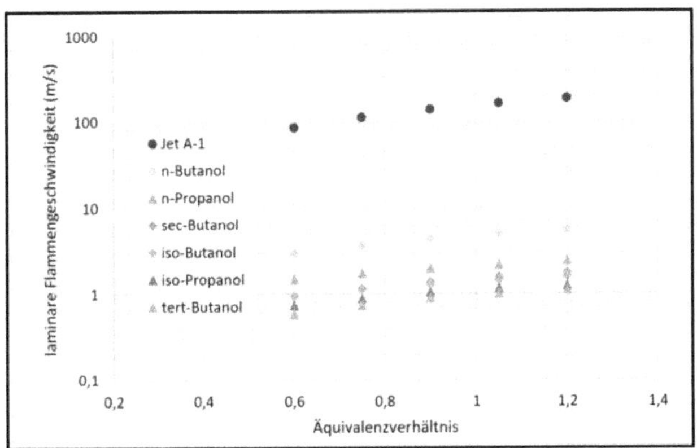

Bild 1: Laminare Flammengeschwindigkeit (berechnet) für Jet A-1 und im Vergleich dazu für C3- und C4-Alkohole für verschiedene Brennstoff-Luft-Gemische bei flugzeugrelevanten Bedingungen (T = 860 K und p = 36 bar) [11].

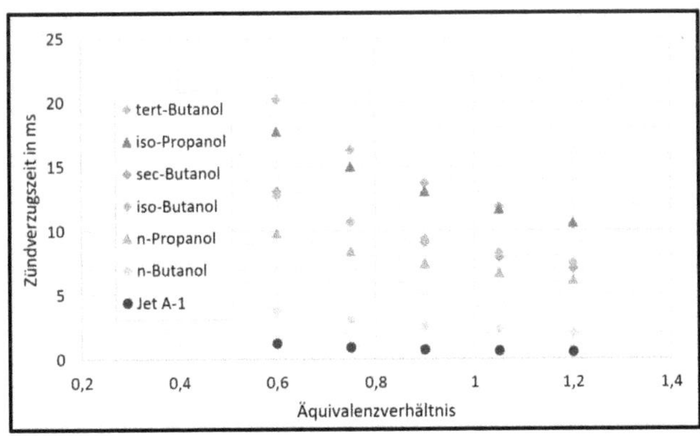

Bild 2: Zündverzugszeit (berechnet) für Jet A-1 und im Vergleich dazu für C3- und C4-Alkohole für verschiedene Brennstoff-Luft-Gemische bei flugzeugrelevanten Bedingungen (T = 860 K und p = 36 bar) [11].

Auch die Neigung zur Rußbildung wurde in Simulationen untersucht. Diese ist bei Propanol am geringsten, bei Butanol höher, aber auch hier deutlich geringer als beim bisherigen Treibstoff Jet A-1. Eine weitere wichtige Eigenschaft ist der Schmelzpunkt, der bei dem verbrennungstechnisch interessanten tert-Butanol allerdings problematisch gering ist, so dass hier der Brennstoff bei üblichen Tankbedingungen einen festen Zustand aufweisen würde [11]. In Tabelle 2 sind zusätzlich auch der Energieinhalt enthalten, sowie Vergleichsdaten für Jet A, Methanol und Ethanol, wobei die letzten beiden doch einen recht niedrigen Energieinhalt haben.

Tabelle 2: Vergleich relevanter Verbrennungsdaten für verschiedene potentielle Brennstoff-Luft Gemische bei flugzeugrelevanten Bedingungen (T = 860 K und p = 36 bar bei einer Luftzahl von 1,5).

	Energy Content LHV (MJ/kg)	**Energy Content** LHV (MJ/l)	**Laminar Burning Velocity** in m/	**Ignition Delay Time** in ms	**Yield Sooting Index**
JetA	43,2	34,9	113,4	0,51	68
Methanol	19,9	15,8	1,21	12,6	-36,9
Ethanol	26,7	22,8	1,14	13,2	-31,1
n-Propanol	30,9	24,7	1,72	7,5	-22
iso-Propanol	30,7	24,1	0,88	13,3	-17
n-Butanol	33,1	26,8	3,64	2,5	-13
sec-Butanol	33,2	26,9	1,15	9,2	-8,6
tert-Butanol (solid)	32,2	25,5	0,73	13,8	-4,5
iso-Butanol	33	26,4	1,11	9,7	-6,5

Im nächsten Forschungsschritt werden wir in einer vorheizbaren vereinfachten LPP-Anordnung die direkte Einmischung verschiedener Brennstoffe und -gemische experimentell untersuchen und hier die Gemischbildung, die Zündgrenzen und auch die Flammenrückschlagsgrenzen genauer untersuchen.

Hierzu wurde eine Versuchsanlage aufgebaut, in der die Vorverdampfung, Vermischung und Flammenstabilisierung für flexibel wählbare Kraftstoffe möglich ist. In einem Vorläuferprojekt wurden hier bereits gasförmige alternative Kraftstoffe untersucht, besonders Ammoniak-Wasserstoff-Gemische, bei denen es besonders um den Flammenstabilitätsbereich ging, also um die Betriebsgrenzen für einen Flammenrückschlag, wenn die Strömungsgeschwindigkeit im Mischrohr zu klein wird, so dass aus dem Brennerbereich die Flamme in diesen Mischbereich zurückschlägt. Bei allen vorgemischten Flammen ist dieses Flashback-Limit sehr wichtig, um einen sicheren Betrieb zu gewährleisten. Goldmann konnte in seiner Dissertation [13] einen sehr breiten Bereich an verschiedenen Ammoniak-Wasserstoff-Gemischen untersuchen. Diese Brennstoffe haben den großen Vorteil, dass keinerlei Kohlenstoff enthalten ist, so dass auch keinerlei CO_2 als Treibhausgas entsteht. Ammoniak (NH_3) ist allerdings einerseits als giftiger Energieträger nicht im Flugzeug gewünscht, und andererseits sehr reaktionsträge, was immerhin durch den sehr reaktionsfreundlichen Wasserstoff gut ausgeglichen werden kann, wenn die richtige Mischung gewählt wird. Bild 3 zeigt den Kern der Versuchsapparatur mit dem Zumischungsbereich mit statischen Mischern zur Homogenisierung unten, einem 500 mm langen Quarzglasrohr mit Durchmesser 20 mm, in dem zudem der Flashback auch optisch untersucht werden kann, und dem Brenner oben. Nicht gezeigt sind die sehr genauen Flow-Controller, die automatische Zündeinrichtung und die vollständig automatisierte Messtechnik. Rechts sind die sehr umfangreichen Messergebnisse aus 351 einzelnen automatisiert durchgeführten Messungen dargestellt [13, 14].

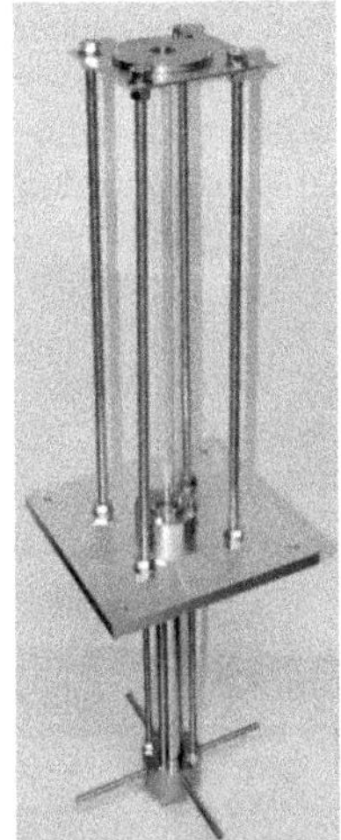

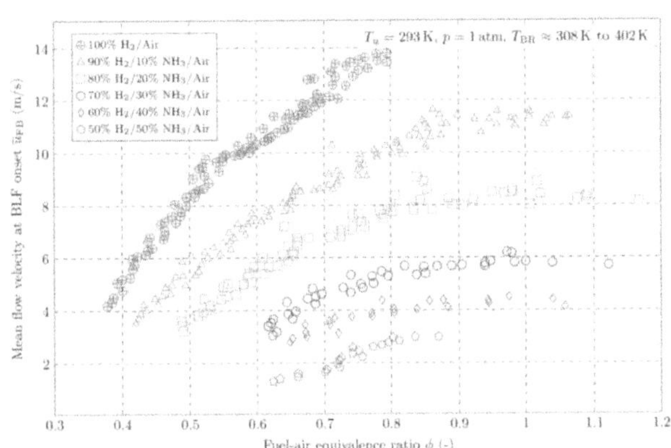

Bild 3: Hauptteil der Versuchsanlage für Mischungs- und Flammenstabilisierungsmessungen (links) und umfangreiche Ergebnisse zum Flammenrückschlags-Limit für Ammoniak-Wasserstoff-Gemische bei verschiedener Brennstoff-Luft-Mischung (rechts) [14].

4. Ausblick und Zusammenfassung

Aktuell wird der Brenner auf flüssige Brennstoffe umgebaut und mit einer Luftvorheizung versehen, so dass hier nun verschiedene potentielle flüssige eFuels erprobt werden können.

In der Planung ist zudem eine auf hohen Druck und hohe Temperatur erweiterte Versuchsanlage, bei der mit bis zu 10 bar und Vorheiztemperaturen bis zu 900 K die Vorverdampfung und Vormischung und die Stabilitätsgrenzen bezüglich Selbstzündung und Flammenrückschlag bei den Bedingungen untersucht werden können, die im realen Flugtriebwerk in Reiseflughöhe herrschen.

Wenn hier erfolgreiche Versuchsergebnisse erreicht werden können, wäre ein wichtiger Schritt im Grundlagenforschungsbereich erreicht, so dass die Vision einer ultrasauberen nachhaltigen Luftfahrt ein Stück näher an die Realisierung gerückt wird. Weitere Schritte sind dann natürlich im Bereich noch höherer Betriebsdrücke und größerer Kraftstoffmengen notwendig, um über skalierte Demonstratoren hin zu realen Flugtriebwerken zu kommen.

Tabelle 3 fasst den Vergleich der besprochenen alternativen Flugantriebskonzepte zusammen. Hier ist zusätzlich auch ein Wasserstoff-Gasturbinen-Antrieb mit aufgeführt. Auch wenn die Abschätzungen des Klima-Einflusses teilweise noch auf weniger genau bekannten Daten beruhen (tw. [5]) zeigt der Vergleich, dass ein Ansatz mit neuen eFuels, sofern der Ansatz der LPP-Verbrennung umsetzbar ist, eine echte Alternative darstellt, weil die Gewichts- und Tankproblematik von Batterie- oder Wasserstoffflugzeugen und die Emissionsproblematik von Kerosin oder SAF-Kerosin hier stark reduziert wäre.

Tabelle 3: Vergleich alternativer Luftfahrt-Antriebskonzepte mit Vor- und Nachteilen. Die Abschätzung des summierten Umwelteinflusses basiert teilweise auf wenig bekannten Daten und ist somit mit Vorsicht zu betrachten.

	Mass Range	Tank (Battery)	Energy-Efficiency Production	Direct GHG Emission	Non-CO2 Emission	Climate Impact (100%=CO2)	Challenges
Kerosene - GT (Basis)	+ +	+ +	fossil	- -	- -	300%	CO2, Emiss.
Battery - El	- -	-	+ +	+ +	+ +	30%	Mass too big
H2 - FC - El	+	-	+	+ +	+ +	50%	Tank, FC
H2 - GT	+	-	+	+ +	+	80%	Tank, NOx
SAF Keros. - GT	+ +	+ +	-	+	-	200%	Emission
eFuel LPP - GT	+	+ +	-	+ +	+ +	50%	to be shown

Es kann angenommen werden, dass die enormen Vorteile der flüssigen Energieträger deren höheren Aufwand bei der Herstellung aus nachhaltigem Strom mehr als wett machen werden. Die potentiell viel sauberere Verbrennung durch die magere Vormischung (LPP-Verfahren), die praktisch rußpartikelfrei und nahezu NO_x-frei verlaufen kann, und die zudem noch kleinere Brennkammern als schon bisher ermöglichen würde, stellt somit eine sehr interessante Option für eine nachhaltige Luftfahrt der Zukunft dar.

Danksagung

Wir danken der DFG für die Förderung im Rahmen des Excellence Clusters EXC 2163/1 "Sustainable and Energy Efficient Aviation" (SE²A), welches von der TU Braunschweig in Zusammenarbeit mit der Leibniz Universität Hannover, der PTB Braunschweig und der DLR Braunschweig und Göttingen durchgeführt wird. WIr danken den Projektpartnern für die sehr gute Zusammenarbeit. Die Forschungsarbeiten am ITV Hannover werden insbesondere von P. Zimmermann durchgeführt, für die Vorarbeiten war A. Goldmann sehr stark involviert. Die Forschungsarbeiten an der PTB Braunschweig werden von R. Fernandes und B. Shu geleitet und von S. Nadiri und L. Sane durchgeführt. U. Schröder (TU Braunschweig) und seine Mitarbeiter waren insbesondere auch an der Auswahl potentieller eFuels beteiligt.

Literatur

[1] *IATA Fact Sheet: Industry Statistics,* https://www.iata.org/en/iata-repository/pressroom/factsheets/industry-statistics/ [Abruf: 15.08.2021].

[2] Europäische Kommission, *European Green Deal: Commission proposes transformation of EU economy and society to meet climate ambitions,* https://ec.europa.eu/commission/presscorner/detail/en/IP_21_3541 [Abruf: 15.08.2021].

[3] Karpuk, S.; Elham, A. *Influence of Novel Airframe Technologies on the Feasibility of Fully Electric Regional Aviation.* Aerospace 2021, 8, 163. DOI: 10.3390/aerospace8060163

[4] Eiselin, S.: *Airbus will 2025 erstmals mit Wasserstofftanks abheben.* In: aeroTelegraph 14. Juni 2021. Abgerufen 04.05.2022.

[5] Lee D.S., Fahey D.W., Skowron A., Allen M.R., Burkhardt U., Chen Q., Doherty S.J., Freeman S., Forster P.M., Fuglestvedt J., Gettelman A., León R.R. de, Lim L.L., Lund M.T., Millar R.J., Owen B., Penner J.E.Pitari G., Prather M.J., Sausen R., Wilcox L.J., *The contribution of global aviation to anthropogenic climate forcing for 2000 to 2018,* Atmospheric environment, 117834, 2021. DOI: 10.1016/j.atmosenv.2020.117834.

[6] Penanhoat O., *Low Emissions Combustor Technology Developments in the European Programmes LOPOCOTEP and TLC,* 25th International Congress of the Aeronautical Sciences, 2006.

[7] Marek C.J., Papathakos, LC: Verbulecz, PW, *Preliminary Studies of Autoignition and Flashback in a Premixing-Prevaporizing Flame Tube Using Jet-A Fuel at Lean Equivalence Ratios,* NASA Technical Memorandum, 1977.

[8] Schäfer O., Koch R., Wittig S., *Flashback in Lean Prevaporized Premixed Combustion: Nonswirling Turbulent Pipe Flow Study,* Journal of Engineering for Gas Turbines and Power, 3, 670–676, 2003. DOI: 10.1115/1.1581897.

[9] Goldmann A., Sauter W., Oettinger M., Kluge T., Schröder U., Seume J., Friedrichs J., Dinkelacker F., *A Study on Electrofuels in Aviation,* Energies, 2, 392, 2018. DOI: 10.3390/en11020392.

[10] Lechner, C., Seume, J., *Stationäre Gasturbinen*, 3. Auflage, Springer Vieweg, 2019.

[11] Zimmermann, P., Nadiri, S., Sane, L., Shu, B., Fernandes, R., Dinkelacker, F., *Numerical investigations on the combustion characteristics of propanol and butanol isomers for sustainable aviation application,* 30. Deutscher Flammentag - für nachhaltige Verbrennung (Hrsg. F. Dinkelacker, H. Pitsch, V. Scherer), Hannover-Garbsen, 28.-29.09.2021, Tagungsband S. 1126-1133, 2021.

[12] Nadiri, S., Zimmermann, P., Sane, L., Fernandes, R., Dinkelacker, F., Shu, B., *Kinetic modeling study on the combustion characterization of synthetic C3 and C4 alcohols for lean premixed prevaporized combustion,* Energies, 2021, 14, 5473. DOI:10.3390/en14175473.

[13] Goldmann, A., *Analyse des Grenzschichtrückschlags von Wasserstoff-Ammoniak Vormischflammen in drallfreien Strömungen*, Dissertation Leibniz Universität Hannover, TEWISS-Verlag Garbsen: Berichte aus dem ITV, Band 2/2021.

[14] Goldmann, A., Dinkelacker, F., *Experimental Investigation and Modeling of Boundary Layer Flashback for Non-Swirling Premixed Hydrogen/Ammonia/Air Flames*, Comb. Flame 226, 362-379, 2021. DOI: 10.1016/j.combustflame.2020.12.021.

Klimaneutrale Mobilität mit „Stroh im Tank“

Birgit Maria Wöber

Der grüne Verbrenner (ICE)

CNG/BioCNG (Compressed Natural Gas), gasförmig komprimiertes Naturgas. Der Methan-Kraftstoff kann sowohl fossil, erneuerbar oder synthetisch hergestellt werden, der Oktanwert liegt bei ~130, bezogen auf H-Gas-Qualität. Die Abgabe-Einheit ist in Deutschland Kilogramm. Der Schwerpunkt des Vortrages bezieht auf die Nutzung von erneuerbarem Biomethan aus Rest- und Abfallstoffen, Kraftstoff „Made in Germany“. Er kann von 0 bis 100 % beliebig beigefügt werden. Gegenüber den „fossilen Flüssigen“ ist er auf einem niedrigeren Preisniveau. Nachhaltiges BioCNG ist ohne technische Veränderung in jedem Fahrzeug mit CNG-Antrieb nutzbar. Der Kraftstoff unterliegt der DIN_EN_16942. Damit lässt sich die europäisch angestrebte Defossilisierung kostengünstig und sofort umsetzen, die Wertschöpfungskette bleibt in Deutschland oder im jeweiligen Land.

Darstellung der Infrastruktur

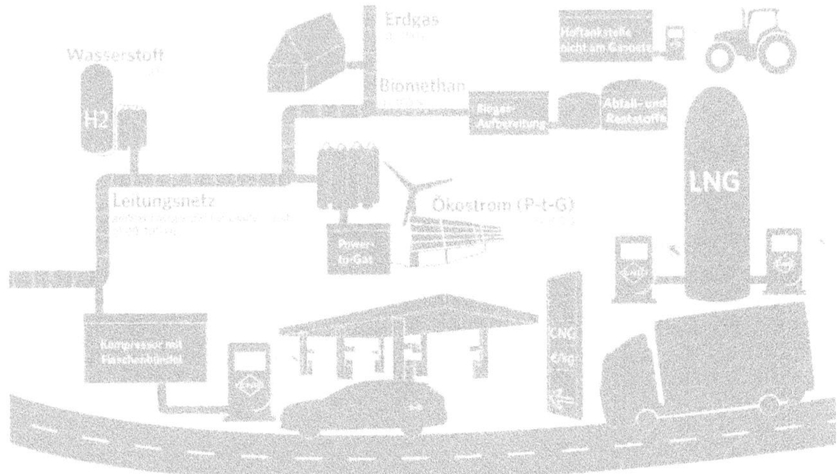

Quelle: CNG-Club e. V.

Methan ist überwiegend ein leitungsgebundener Kraftstoff. Er nutzt das weitverzweigte Pipeline-Netz in Deutschland von über 500.000 km. Biomethan kann an Einspeisepunkten dem Leitungsnetz zugeführt werden, ebenso ist die Entnahme durch das Tanken möglich. Ab dem Zeitpunkt der Komprimierung spricht man von CNG. Zukünftig kann der Methankraftstoff jedoch auch durch leitungsunabhängige CNG-Tankstellen an größeren Biogasanlagen und auch an LNG-Tankstellen angeboten werden.

Biomethananteil im CNG

Entsprechend der obigen Beschreibung der Infrastruktur lässt sich die Defossilisierung und die damit verbundene Erreichung von Klimaschutzzielen leicht umsetzen. Die Entwicklung der Biomethananteile hat sich im Zeitfenster von 2018 bis 2022 erfreulich positiv entwickelt.

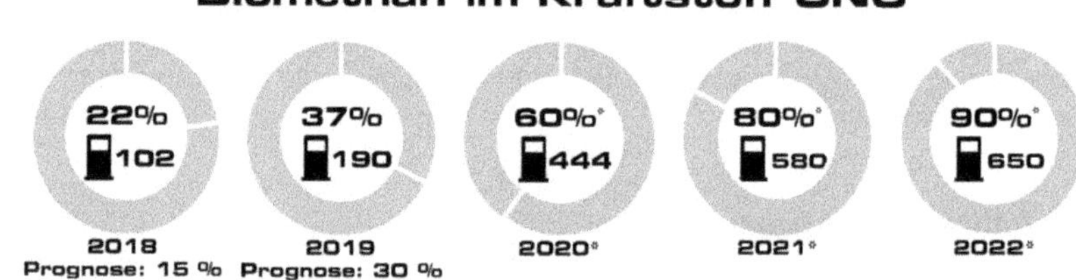

Quelle: CNG-Club e. V.

Die Vorteile liegen auf der Hand: Konstant erneuerbar, egal ob der Wind weht oder die Sonne scheint. Biomasse ist ausreichend vorhanden. BioCNG zählt zu den fortschrittlichen Kraftstoffen aus Rest- und Abfallstoffen wie Stroh, Schlempe, Gülle, organische Abfälle. Durch Gülle werden Negativ-CO_2-Werte bei der THG-Bewertung möglich. Aktuell sind in Deutschland 785 CNG Tankstellen (Stand 04/2022) in Betrieb.

Ausgangsmaterialien für die Biomethanerzeugung

Die Grafik zeigt eine Auswertung einer CNG-Tankstelle in Bayern. Rückwirkend auf das Jahr 2021 kann konkret nachgewiesen werden, welche Rest- und Abfallstoffe an dieser Station vertankt wurden und wieviel CO_2-Reduktion damit erreicht wurde.

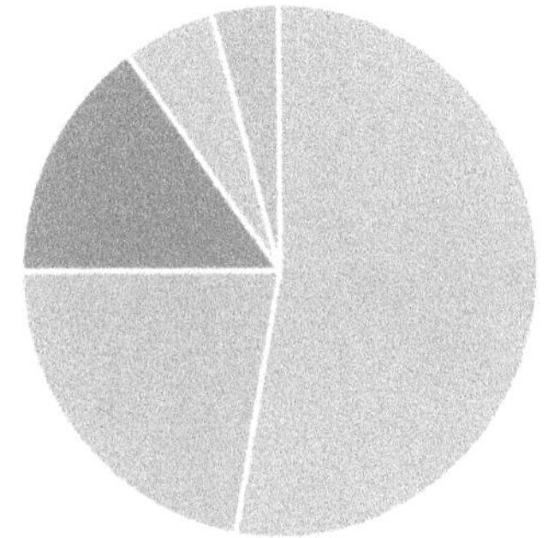

Quelle: CNG-Club e. V.

Vorstellung einer Biomethanproduktion am Beispiel der VERBIO AG

VERBIO ist ein führender Bioenergie-Produzent in Europa. VERBIO produziert Biokraftstoffe, Futtermittel, Biodünger und Rohstoffe für die Pharma-, Kosmetik- und Nahrungsmittelindustrie. VERBIO betreibt Tochtergesellschaften in Polen, Ungarn, Indien, USA und Canada. VERBIO produziert Biodiesel, Bioethanol und Biomethan mit einem Marktanteil von ca. 20 %. Verbio ist weltgrößter Biomethanproduzent. Weitere Infos auf www.verbio.de und www.verbiogas.de.

Wie kommt das Stroh in den Tank? Dank Kreislaufwirtschaft!

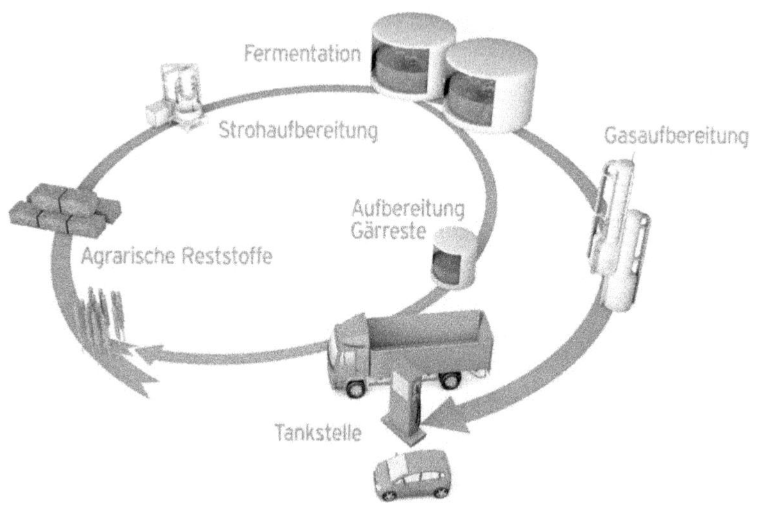

Quelle: Verbio AG

Fuel Science: Molecularly Controlled Combustion with Bio-hybrid Fuels

Bastian Lehrheuer, Patrick Burkardt, Christian Honecker,
Maximilian Fleischmann, Stefan Pischinger

Kurzfassung

Eine der größten gesellschaftlichen Herausforderungen ist, trotz steigender Energienachfrage, die derzeitige fossile Energieversorgung durch regenerative Ressourcen zu ersetzen, um dem Klimawandel entgegenzuwirken. Die zunehmende Verfügbarkeit nicht-fossiler Energietechnologien eröffnet ungeahnte Möglichkeiten, die Schnittstelle zwischen Energie- und Stoffwertschöpfungsketten für eine nachhaltige Zukunft neu zu gestalten. Der Exzellenzcluster 2186 „The Fuel Science Center" (FSC) bringt daher Forscher aus verschiedenen Disziplinen zusammen, um das heutige statische, auf fossilen Brennstoffen basierende Szenario durch adaptive Produktions- und Antriebssysteme zu ersetzen, die auf erneuerbaren Energie- und Kohlenstoffressourcen unter dynamischen Systemgrenzen basieren. Ziel der Arbeiten von FSC ist die Integration von erneuerbarem Strom mit der gemeinsamen Nutzung von biobasierten Kohlenstoffrohstoffen und CO_2, um flüssige Kraftstoffe mit hoher Energiedichte ("Bio-hybrid Fuels") bereitzustellen, die innovative Motorkonzepte für eine hocheffiziente und saubere Verbrennung ermöglichen. Eine Schlüsselkomponente der Forschung ist der "Fuel Design Process", der einen integrierten Ansatz zwischen Kraftstoffsynthese und Antriebstechnik darstellt. Am Lehrstuhl für Thermodynamik mobiler Energiewandlungssysteme (TME) werden diese Bio-hybrid Fuels und deren Mischungen in neuartigen, molekular gesteuerten Brennverfahren, den Molecularly Controlled Combustion Systems (MCCS), erforscht.

1. Research Outline

One of the greatest challenges facing society is to replace the current fossil energy supply with renewable resources in order to counteract climate change, despite an increasing demand for energy. The growing availability of non-fossil energy technologies opens up unprecedented opportunities to redesign the interface between energy and material value chains for a sustainable future. The Fuel Science Center (FSC) therefore brings together researchers from different disciplines to replace today's static fossil fuel-based scenario with adaptive production and propulsion systems based on renewable energies and carbon resources under dynamic system boundaries. The aim of FSC's work is to integrate renewable electricity with the shared use of bio-based carbon raw materials and CO_2 to provide high energy-density liquid fuels ("Bio-hybrid

Fuels") that enable innovative engine concepts for highly efficient and clean combustion. A key component of the research is the "Fuel Design Process", which represents an integrated approach between fuel synthesis and propulsion technology.
At the Chair of Thermodynamics of Mobile Energy Conversion Systems (TME), these Bio-hybrid Fuels and their mixtures are being researched in novel, molecularly controlled combustion systems (MCCS). On a compression ignition engine, in-cylinder fuel blending was investigated to evaluate the required fuel reactivity for a Molecular Spark Combustion System (Figure 1). For a first proof of concept, ethanol was used as homogeneous low reactivity fuel combined with various octanol-ethanol blends as stratified high reactivity fuel. On the spark-ignition counterpart, the in-cylinder fuel blending was realized using an active pre-chamber application, the Molecular Torch (Figure 3), with a combination of methanol injected into the main combustion chamber and an ethanol-fueled pre-chamber. In addition to the engine investigations, the pre-chamber combustion system was optically investigated on a rapid compression machine. One of the key findings is the visualization of the combustion behavior in and outside the pre-chamber with novel bio-hybrid fuels under engine conditions.

2. Molecular Spark in Compression Ignition Engines

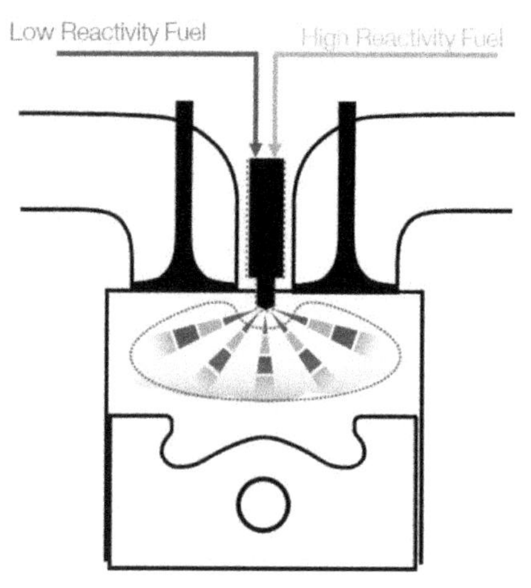

Figure 1: Molecular Spark

To determine the required steps to achieve a very high efficiency with simultaneously strongly reduced pollutant emissions, a simulative, theoretical assessment of influencing factors has been performed [1]. The heat-release-shape and compression ratio were identified as the most relevant parameters to gain efficiency improvements. Initial experiments were carried out in TME's light duty V_h = 390 cm³ single cylinder diesel research engine, utilizing ethanol as low-reactivity fuel and 1-octanol as high-reactivity fuel, using a piston with a conventional omega-bowl and a compression ratio (CR) of CR = 19.

In order to utilize the full efficiency potential, the engine must be operated with a high compression ratio, which poses a challenge for dual fuel systems due to knocking combustion. The molecularly controlled combustion concept relies on long ignition delays to shape and distribute the molecular spark in a manner that allows for fast heat-release and knock-mitigation simultaneously. Under higher compression ratios, ignition delays are reduced and knocking tendencies are increased considerably.

To operate with a compression ratio of CR = 19:1 instead of the initial CR = 15:1, high rates of exhaust gas recirculation (EGR) are used to lower charge reactivity, thus mitigating knocking tendencies. Yet, for classical diesel fuels as high reactivity fuel, knock remains challenging. Later centers of combustion and lower shares of high reactivity fuel lead to deteriorated combustion stability and efficiency as the molecular spark cannot form as intended. However, less reactive high-reactivity fuels perform better under the increased compression ratio, as their ignition delays under the resulting high pressures and temperatures approach the desired values.

It was found beneficial to provide a long ignition delay time, just at the turnover point to reactivity control, to allow the high reactivity fuel to distribute throughout a large part of the combustion chamber, evaporate fully and largely mix with the stratified, highly lean charge of low reactivity fuel. In combination with an approximately 50:50-share in energy content between the two fuels, this allows for the formation of a comparably large molecular spark, i.e. a zone in the cylinder in which reactivity suffices for ignition. This molecular spark then serves as a strong ignition source, thus achieving stable and relatively complete combustion. Most of the heat of combustion is released in the fast heat-release regime and is thus available for generation of expansion work for most of the piston-stroke, thereby yielding efficiency benefits. As peak heat release rates are, however, in the range of conventional diesel combustion, knocking in the stratified lean charge of low-reactivity fuel can be averted.

A screening of high-reactivity fuels was carried out utilizing 1-octanol blended with 0, 10 and 30 vol.-% of ethanol, resulting in derived cetane numbers (DCN) of 34.8, 30.5, and 17.5, respectively (see Figure 2) [2]. The high reactivity fuel mixture containing 10 vol.-% Ethanol and thereby a derived cetane number of about 30 was found to be most suitable to achieve a high indicated efficiency ($\eta_{i,net}$ = 48.8 %) with lowest pollutant emissions. Based on the results of this screening, hexyl acetate, which has a cetane number of 28, was found to be a promising Bio-hybrid Fuel candidate for the high-reactivity fuel. In the following tests, these expectations could be validated, with hexyl acetate performing well under the desired conditions and achieving an indicated efficiency of up to $\eta_{i,net}$ = 48.8 % at an engine speed of n = 2000 1/min and an indicated mean effective pressure of IMEP = 10 bar.

3. Molecular Torch in Spark Ignition Engines

The alternative alcohol neat fuels methanol and ethanol were thermodynamically investigated on a direct-injection spark-ignition single-cylinder engine for passenger car applications with a compression ratio of CR = 16.4 [3,4]. However, instead of a conventional spark ignition (SI) combustion system, the engine was equipped with an active pre-chamber (PC) combustion system. In particular, the PC was fueled with ethanol, while both methanol and ethanol were investigated as the main combustion chamber (MC) fuels.

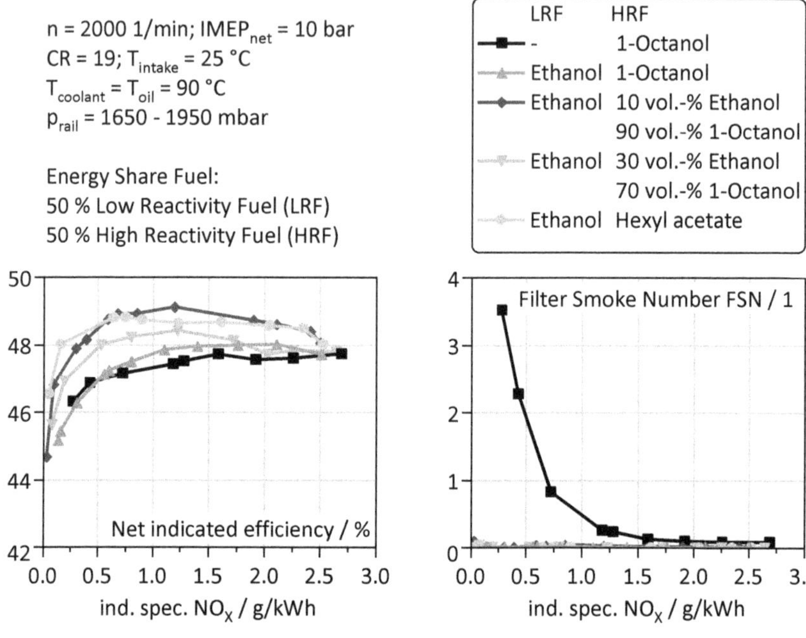

Figure 2: Variation of High Reactivity Fuel (HRF) Composition to determine its optimum characteristics [1]

The fuels were assessed in a variation of the relative air/fuel ratio (λ) and the results were compared to those of a conventional SI combustion system fueled with methanol respectively ethanol.

The PC applications achieved higher lean limits than their SI counterparts (see Figure 2). This possibility resulted from the enriched air/fuel mixture in the PC due to the additional fuel injection into the PC. Thus, the mixture in the PC can be maintained within the ignition limits at high global λ. When comparing different MC fuels, methanol achieved higher lean limits than ethanol in the combustion systems of both SI and PC. Here, metha-

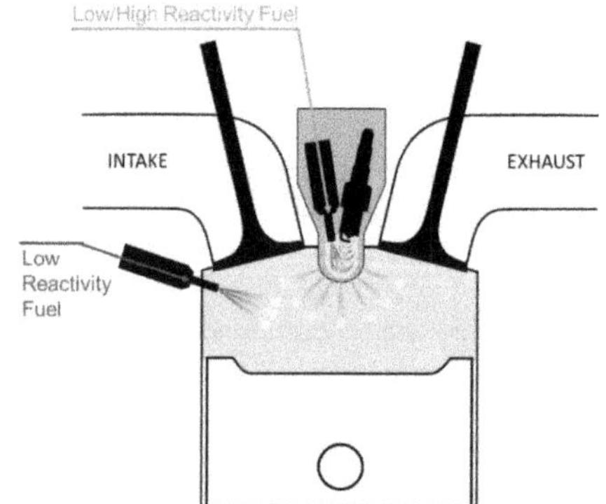

Figure 3: Molecular Torch

nol's higher laminar flame speed compared to that of ethanol contributed to the achieved lean limits.

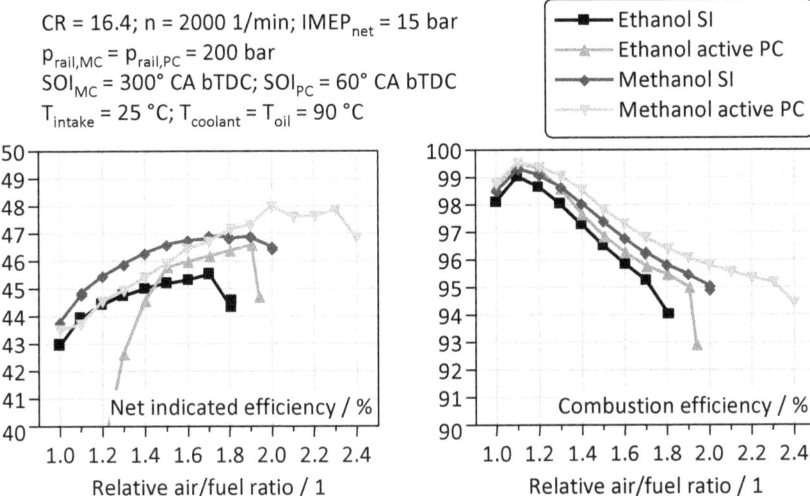

Figure 4: Results of a relative air/fuel ratio variation with both methanol and ethanol in a conventional spark-ignition combustion system and in a pre-chamber combustion system. [3,4]

With respect to $\eta_{i,net}$, the SI combustion systems were favorable at low λ, while the PC combustion systems achieved higher values at high λ. At low λ, the losses of a PC combustion system are pronounced: first, the additional volume and the high surface/volume ratio result in increased wall heat losses. Second, the flow losses due to the PC orifices are an additional efficiency loss compared to a conventional SI layout. Third, the fuel fraction supplied from the MC and the fuel directly injected into the PC do not entirely contribute to an increase in power. All these losses typical for PC combustion systems resulted in a lower $\eta_{i,net}$ in case of the PC system compared to the SI system. However, these losses became less pronounced at high λ. In addition, the PC combustion systems open the opportunity to operate a higher λ than the SI counterparts. Thus, the PC combustion system can overcompensate for its losses and achieve higher $\eta_{i,net}$ than its SI counterpart. In particular, a maximum of $\eta_{i,net}$ = 48.0 % was achieved with the PC combustion system with methanol-fueled MC at λ = 2.0.

3.1 Optical Pre-chamber Experiments

Optical investigations of the molecularly controlled combustion inside a rapid compression machine (RCM) were performed to gain fundamental knowledge of the pre-chamber combustion for engine conditions [4]. Good optical access and well-defined boundary conditions were the main reasons for using the RCM for the analysis of the active pre-chamber combustion. The first optical results showed comparably low natural luminosity of the FSC fuels inside the visible light. Therefore, a sodium tracer-fuel mixture was injected into the pre-chamber for higher light intensity and pre-chamber jet visualization. The comparison of experiments with and without tracer showed an insignificant influence of the sodium tracer on the pre-chamber combustion behavior.

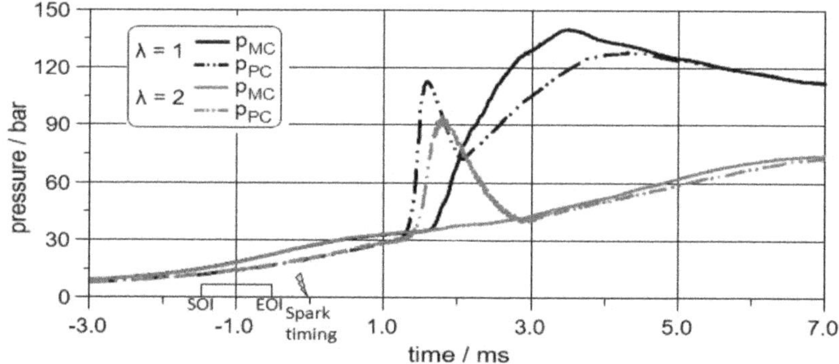

Figure 5: RCM measurements for DEM and ethyl acetate with λ_{MC} variation.

The frame rate for the high-speed measurements in the active pre-chamber was 4000 frames per second. The corresponding pressure is plotted in Figure 4 in black color. Figure 5 shows the first proof of concept to visualize the injection, spark discharge, and combustion inside an originally scaled pre-chamber with a volume of V_{pc} = 1380 mm³. The RCM allows simultaneous optical investigation of the combustion inside and outside the pre-chamber to analyze the PC-MC interaction.

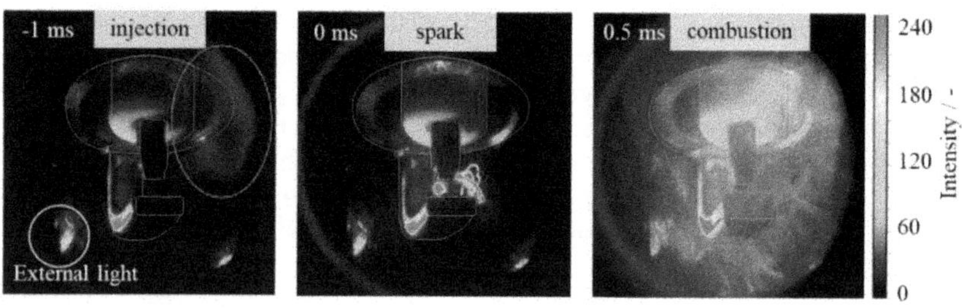

Figure 6: Optical view of natural luminosity for injection, spark discharge, and combustion inside the active pre-chamber.

In a next step, different low and high reactivity FSC fuels will be investigated in the molecularly controlled combustion system with active pre-chamber inside the RCM.

Furthermore, different operating conditions as well as pre-chamber layouts will be investigated to analyze their influence on the PC-MC interaction.

4. Conclusion & Outlook

Starting from today's fuel specific swirl-type compression ignition and tumble-type spark ignition concepts, novel molecularly controlled combustion systems enabled by bio-hybrid fuels show a great potential to improve efficiency and reduce pollutant emissions at the same time. The development of innovative measurement techniques and model-based methodologies integrated in one Fuel Design Process, help to identify the overall optimum of fuel and combustion characteristics. In future work, limitations of the proposed lean burn concepts regarding high load capability, efficiency, stability and unburned HC and CO emissions shall be overcome through combination of the targeted use of specific multiple fuels with advanced ignition concepts, such as the molecular spark and the molecular torch, optimized cylinder flow, catalytic and thermal coatings, and electrically activated aftertreatment systems.

Acknowledgement

This research work is funded by the Deutsche Forschungsgemeinschaft (DFG, German Research Foundation) under Germany´s Excellence Strategy – Exzellenzcluster 2186 „The Fuel Science Center“ ID: 390919832.

WISSENSCHAFTSRAT

Literature

[1] Christian Honecker, and Stefan Pischinger. “Molecularly-Controlled High Swirl Combustion System”. 9th International Conference on Fuel Science “From Production to Propulsion”, Aachen, Germany. June 22 to 24, 2021

[2] Marcel Neumann, Philipp Ackermann, Jan G. Rittig, Christian Honecker, Alexander Mitsos, Manuel Dahmen, and Stefan Pischinger. “Towards Characteristic Fuel Numbers for Molecularly-Controlled Combustion Systems”. 10th International Conference on Fuel Science “From Production to Propulsion”, Aachen, Germany, May 10 to 12, 2022

[3] Patrick Burkardt, Christian Wouters, and Stefan Pischinger. "Potential of alcohol fuels in active and passive pre-chamber applications in a passenger car spark-ignition engine." International Journal of Engine Research, 2021

[4] Patrick Burkardt et al. "On the Use of Active Pre-chambers and Bio-hybrid Fuels in Internal Combustion Engines." Engines and Fuels for Future Transport. Springer, Singapore, 2022. 205-231.

Optical investigations on mixture formation, self-ignition and soot formation of polyoxymethylene-dimethyl-ether (OME)

Lukas Weiß

Kurzfassung

Diese Arbeit steht im Kontext einer Entwicklung von Technologiekonzepten zur Darstellung einer nachhaltigen und klimaneutralen Mobilität. Um diese EU-weit bis 2050 umzusetzen, muss ein signifikanter Beitrag auch durch e-Fuels, aus erneuerbaren Energien synthetisch hergestellten Kraftstoffen, erfolgen. Einer der interessanten Vertreter von e-Fuels ist OME, ein organisches Kettenmolekül, dessen Kettenstruktur alternierend aus Kohlenstoff und Sauerstoff aufgebaut ist und keine Nebenketten hat. In ersten Kraftstoff-Studien ist OME durch seine besonders rußarme Verbrennung im Vergleich zu anderen sauerstoffreichen Kraftstoffen aufgefallen. Daher wird in dieser Arbeit die Prozesskette der OME-Verbrennung unter motorischen Randbedingungen am Beispiel von OME_{3-6} untersucht (der Index gibt die Kettenlänge des Moleküls an). Im Detail wird auf die Einspritzung und Gemischbildung, die Zündung sowie die Rußbildung im Vergleich zu Dodekan und GtL-Diesel eingegangen. Die Fragestellungen der Untersuchung sind, inwieweit die Prozesse von OME vergleichbar zu konventionellen Kraftstoffen sind und wodurch die geringe Rußbildung erklärt werden kann. Während andere Kraftstoffe mit ähnlichem Sauerstoffanteil wie OME in der Literatur sehr geringe, aber dennoch vorhandene Rußbildung zeigen, sind im Fall von OME keine Anzeichen von Ruß in der primären Verbrennung erkennbar. Diese Aussage basiert nicht nur darauf, dass keine thermische Rußstrahlung detektiert werden kann, sondern auch darauf, dass weder PAH als Rußvorläufer noch der elementare Grundbaustein CH nachweisbar ist. Diese Erkenntnis sagt, dass generell bei sauerstoffhaltigen Kraftstoffen vielmehr die vorhandene C-O Bindung als der lokal niedrige Sauerstoffbedarf zur geringen Rußbildung beiträgt.

1. Introduction

This work is in the context of a development of technology concepts for the representation of a sustainable and climate-neutral mobility. While in politics as well as in society there is a strong focus on purely battery-electric drives, the study "E-Fuels - The potential of electricity-based fuels for low emission transportation in the EU" by the German Energy Agency and Ludwig-Bölkow-Systemtechnik (LBST) from 2017 shows that it is not possible to implement climate-neutral mobility in the EU by 2050. This study points out that a significant contribution must also be made by e-Fuels, fuels synthetically produced from renewable energies. [1–3].

One of the interesting representatives of e-Fuels is OME. The abbreviation stands for polyoxy-methylenedimethylether [$CH_3O-(CH_2O)_n-CH_3$], an organic chain molecule

whose chain structure alternates between carbon and oxygen and has no side chains. This fuel can be produced relatively easily. The process route starts with water electrolysis. The hydrogen obtained is synthesized with carbon dioxide to form methanol. This basic chemical can eventually be further processed to OME [4, 5].
In initial fuel studies, OME has attracted attention due to its particularly low-soot combustion compared to other oxygen-rich fuels [6, 7]. Therefore, the process chain of OME combustion under engine boundary conditions is investigated in this work using the example of OME_{3-6} (the index indicates the chain length of the molecule). In detail, the injection and mixture formation, the ignition as well as the soot formation in comparison to dodecane and Gas-to-liquid (GtL) diesel are dealt with. The question of the investigation is to what extent the processes of OME are comparable to conventional fuels and how the low soot formation can be explained.

2. Experimental overview

The tests are carried out on an optical high-pressure combustion test rig (OptiVeP). The ambient pressure is kept constant at 60 bar and the ambient temperature is varied between 600 °C and 700 °C. A scientific reference spray is tested at 1200 bar injection pressure from a Continental PCRs2 injection nozzle for passenger car diesel engines with a hole diameter of 115 µm.
Data acquisition is carried out by means of various optical measurement techniques, most of which are imaging-based and allow high temporal resolutions of the transient spray processes. The measurement techniques interlock to map the individual process steps from mixture formation to combustion. For the mixture formation analysis, injection rate measurements, Mie scattered light and Schlieren images as well as Raman data are used. In the combustion phase, planar laser-induced fluorescence at 355 nm, Schlieren, chemiluminescence images and flame spectroscopy are used. More details on the setups can be found in [8].
The spray propagation of OME sprays can be described equivalently to classical diesel sprays with the momentum balance of Abramovich and the spray model of Musculus and Kattke derived from it.

3. Findings

The momentum balance shows that the comparatively high density of OME leads to lower internal nozzle flow velocities than with conventional fuels. This fact may be important for avoiding cavitation effects in the nozzle.
The similar mass distribution in the spray compared to dodecane or GtL diesel and the simultaneously low calorific value of OME result in very lean fuel-air mixtures, which, depending on the operating point, lead to stoichiometric air numbers smaller than λ = 2 at the lift-off length, see Figure 1.

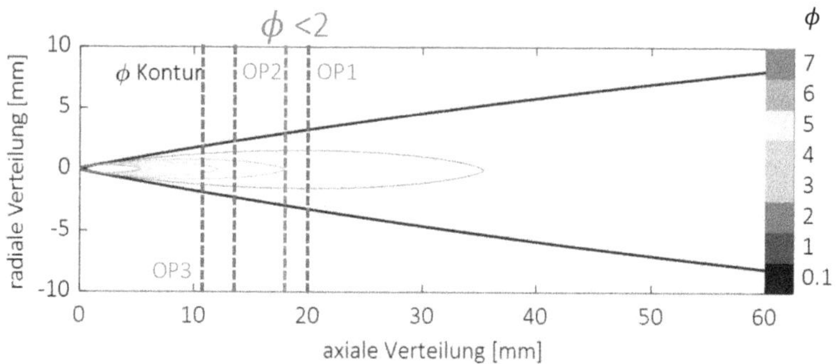

Figure 1: Very low stoichiometric equivalence rations in an OME spray.
Φ <2 at the flame Lift-Off at OP1 and between 3-4 at the high temperature OP3.

The lean mixture results in a high proportion of premixed combustion. This takes place close to the nozzle on the spray axis, resulting in a high energy conversion there. This means that hot combustion gases in conventional diesel engine geometries are expected to hit the pistons in the form of an impingement flow, leading to high thermal stresses on the piston material at the point of impact.

In contrast to conventional diesel engine combustion with conventional fuels, diffusive combustion plays no role in OME conversion. At operating point 1 with an ambient temperature of 600 °C, no signs of diffusive combustion are discernible. When the temperature is increased to 700 °C, small diffusive components are present.

While other fuels with similar oxygen content as OME show very low, but nevertheless present soot formation in the literature [9, 10], in the case of OME no signs of soot in the primary combustion are recognizable. This statement is based not only on the fact that no thermal soot radiation can be detected, but also on the fact that neither PAH as a soot precursor in 355-nm LIF imaging nor the elemental basic building block CH* is detectable, see Figure 2.

The reason for this is that the chemical structure of OME is built up of alternating C and O atoms, does not contain any branches and therefore no C-C bonds occur. Due to their polarity, the C-O bonds are so strong that they are not broken in the combustion process. Thus, classical soot-forming mechanisms are already prevented at the origin. This finding states:

- That secondary sources of soot must be responsible for the positive particulate measurements in exhaust measurements of OME-fired engine test benches.
- That, in general, in the case of oxygenated fuels, the existing C-O bond rather than the locally low oxygen demand contributes to the low soot formation.

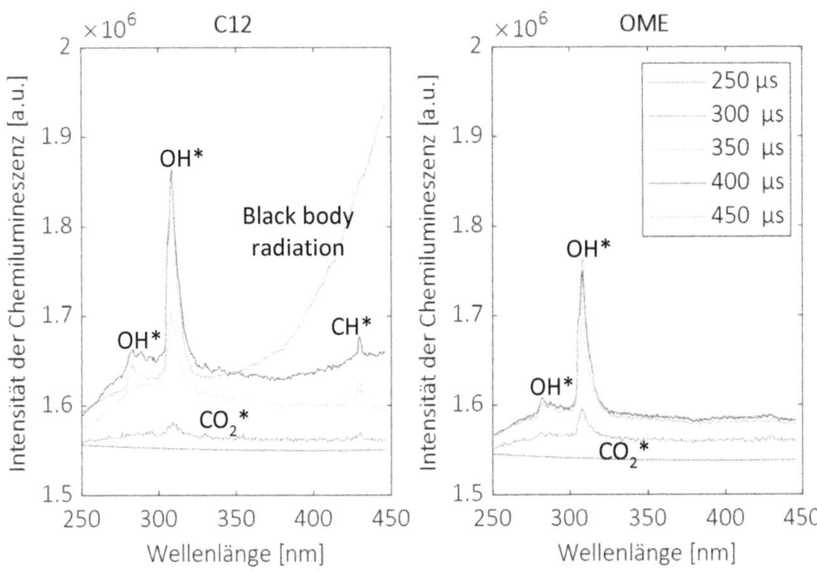

Figure 2: Spectra of the chemiluminescence of a dodecane flame (left) and an OME3 flame (right) at OP3 (700°C) at different times after the start of injection. Black body radiation and CH are not present in the OME flame.*

The latter in particular opens up completely new possibilities for the development of an OME combustion process that no longer needs to have anything in common with a classic diesel combustion process. Thus, relatively small engines in lean-burn operation with a lot of EGR could be used to represent the small to medium load ranges of a vehicle. In this case, EGR brings NO_x emissions to a negligibly low level. For the rare high load cases, a stoichiometric combustion process could be used, which would allow enormous power densities, and, at the same time, a comparatively inexpensive NO_x storage catalytic converter could take over the exhaust gas aftertreatment.

The injection system technology can basically be built using existing common-rail technology. Only seals must be selected based on OME-stable materials. The somewhat lower viscosity requires slight adjustments to the injector internal flows in order to be able to represent the required needle dynamics. The compensation of the low caloric value can be easily realized by increased hole diameters. The advantages of small holes in the classic diesel engine do not come into play in the OME application due to generally low liquid penetration depths and suppressed soot formation.

The piston geometries and piston materials must be viewed critically because of the expected high thermal loads during flame-wall interaction. This is where the greatest need for research still exists. It may be sufficient to adapt the flow guidance in such a way that classic wall impingement flow no longer occurs, but instead the fuel spray is

gently deflected by the piston geometry to capture the space. Because of the absence of soot formation in an OME engine, the high-pulse spray-wall interaction that is important in diesel engines for soot prevention should also no longer play a role, so that alternatively very broadly spread fuel jets offer another possibility for combustion chamber air capture.
Finally, another exciting area of research will be the investigation of multiple injections to control the combustion process in terms of power delivery and noise emissions.

From a current technical perspective, OME fuel represents an interesting opportunity to make a significant contribution to climate-neutral mobility with novel engine concepts. It is not a question of this fuel preserving the classic internal combustion engine out of nostalgia, but rather it offers an important alternative in the future bouquet of drive concepts. The high energy storage density compared to hydrogen or lithium-ion batteries is a decisive advantage for all mobility concepts in which payload, total weight and range are key parameters. The liquid aggregate state and its chemical stability allow easy transport in already existing infrastructure. Additionally, it offers an excellent energy buffer solution in the overall system of a climate-neutral energy supply based on photovoltaics and wind power. Decisive for the market implementation of such a fuel is the legislative recognition of the CO_2 neutrality of the fuel and thus the political will for a technology-open transformation of the energy supply.

References

[1] United Nations, "Summary of the Paris Agreement," *United Nations Framework Convention on Climate Change*, pp. 27–52, 2015, [Online]. Available: http://bigpicture.unfccc.int/#content-the-paris-agreement

[2] N. Regis *et al.*, "Electrification of the Transport System: Studies and Reports," *Renewable and Sustainable Energy Reviews*, vol. 10, no. 6, pp. 1–49, 2017, doi: 10.1016/j.rser.2016.05.025.

[3] S. Siegemund *et al.*, "The potential of electricity-based fuels for low-emission transport in the EU An expertise by LBST and dena," 2017.

[4] N. Schmitz, J. Burger, E. Ströfer, and H. Hasse, "From methanol to the oxygenated diesel fuel poly(oxymethylene) dimethyl ether: An assessment of the production costs," *Fuel*, vol. 185, pp. 67–72, 2016, doi: 10.1016/j.fuel.2016.07.085.

[5] J. Burger, M. Siegert, E. Ströfer, and H. Hasse, "Poly(oxymethylene) dimethyl ethers as components of tailored diesel fuel: Properties, synthesis and purification concepts," *Fuel*, vol. 89, no. 11, pp. 3315–3319, 2010, doi: 10.1016/j.fuel.2010.05.014.

[6] G. Richter and H. Zellbeck, “OME als Kraftstoffersatz im Pkw-Dieselmotor,” *MTZ - Motortechnische Zeitschrift*, vol. 78, no. 12, pp. 66–73, 2017, doi: 10.1007/s35146-017-0131-y.

[7] M. Härtl, P. Seidenspinner, E. Jacob, and G. Wachtmeister, “Oxygenate screening on a heavy-duty diesel engine and emission characteristics of highly oxygenated oxymethylene ether fuel OME1,” *Fuel*, vol. 153, pp. 328–335, 2015, doi: 10.1016/j.fuel.2015.03.012.

[8] Lukas Weiss, “Optische Untersuchungen zur Gemischbildung, Selbstzündung und Rußbildung von Polyoxymethylendimethylether (OME),” Dissertation, Friedrich-Alexander-Universität Erlangen-Nürnberg, Erlangen, 2022.

[9] A. S. (Ed) Cheng, R. W. Dibble, and B. A. Buchholz, “The Effect of Oxygenates on Diesel Engine Particulate Matter,” *SAE Technical Paper*, no. 2002-01–1705, 2002, doi: 10.3109/9781420086461-6.

[10] J. Manin, S. Skeen, and L. Pickett, “Effects of Oxygenated Fuels on Combustion and Soot Formation / Oxidation Processes,” *SAE International Journal of Fuels and Lubricants*, vol. 7, no. 3, pp. 704–717, 2014.

Methanol to Gasoline: Ein effizienter Weg zur Senkung der Treibhausgasemissionen in der bestehenden Fahrzeugflotte

Thomas Garbe, William Kaszas, Martin Schüttenhelm

Abstract

The MtG process is suitable for producing synthetic fuel components, which are largely the same as conventional base fuel. Since it is possible to produce these components regeneratively, the use of this process can provide a fuel that is new to greenhouse gases. The use of such fuels is possible in the vehicle fleet, so there is a short-term option for greenhouse gas-neutral mobility with existing gasoline vehicles.
As part of the C3-Mobility project, Volkswagen carried out emission tests with MTG blends containing 10 and 20 percent ethanol, respectively. In various emission tests, it has been shown that when using these blends, the pollutant and CO_2 emissions are in the order of magnitude of operation with conventional EN 228-compliant fuels. It has also been shown that an application to the E20 fuel with a higher compression allows slight advantages of about 3.5% CO_2. A fuel with improved fossil raw materials was also measured on a vehicle modified in this way. It is characterized by a high octane number and a low content of heavy aromatics, which contribute particularly to particle formation. This vehicle-fuel combination shows particle reductions of over 70% compared to the starting fuel in the standard engine and a CO_2 reduction of over 5%.
It is recommended to combine a rapid scaling of the MtG process with a simultaneous improvement of the basic quality.

1. Einleitung

Aus der notwendigen Begrenzung der Erderwärmung resultiert nicht nur ein Zeitpunkt, an welchem die Netto-Treibhausgasemission auf Null abgesenkt werden, sondern auch eine bis dahin maximal zu emittierende Menge an Treibhausgasen. Dieses so genannte Treibhausgasbudget ist derart niedrig, dass alle Emittenten umgehend signifikant zur Reduzierung der Treibhausgasemissionen beitragen müssen. Üblicherweise wird deshalb eine sektorbezogene Betrachtung vorgenommen, um Transparenz zu erzeugen, und geeignete Maßnahmenkataloge abzuleiten.
Im Verkehrssektor ist der schnelle und umfassende Einsatz elektrischer Fahrzeuge der wirksamste Beitrag zum Klimaschutz und somit zwingend notwendig. Bis 2050 kann der gesamte Fahrzeugbestand auf Elektroantriebe umgestellt werden und mit regenerativem Strom angetrieben werden. Allerdings sind weltweit derzeit noch über eine Milliarde Fahrzeuge mit Verbrennungsmotor in Betrieb [1]. Der Aufbau der Produktion von Elektrofahrzeugen ebenso wie der Ausbau der Erzeugung von Grünstrom

und einer geeigneten Ladeinfrastruktur erfolgen in ambitionierter Weise. Dennoch werden aktuell noch deutlich mehr Fahrzeuge mit Verbrennungsmotor in Betrieb genommen als Elektrofahrzeuge. Insofern ist davon auszugehen, dass auch in den 30er Jahren noch eine erhebliche Menge an Fahrzeugen mit Verbrennungsmotor auf den Straßen dieser Welt unterwegs sein wird. Nach heutiger Einschätzung muss aber zu dieser Zeit die Phase der Treibhausgasneutralität erreicht sein. Es ist also zwingend notwendig, für diese Fahrzeuge eine Möglichkeit zu schaffen, diese treibgasreduziert oder im Idealfall treibhausgasneutral zu betreiben.
Volkswagen hat gemeinsam mit Partnern eine Phase-In Strategie für regenerative Kraftstoffe entwickelt, welche als ersten Schritt das Blending abgestimmter Komponenten mit fossiler Grundware vorsieht. Um Transparenz für den Kunden zu generieren, wurde eine „R“ Kennzeichnung etabliert, welche den regenerativen Anteil eines Kraftstoffes bezeichnet. Die in Bild 1 dargestellte Kraftstoffstrategie zeigt schematisch den Einsatz neuer Kraftstoffe als notwendige Hauptsorte; lokal können durchaus heute schon Kraftstoffe mit höheren regenerativen Anteilen verfügbar sein.

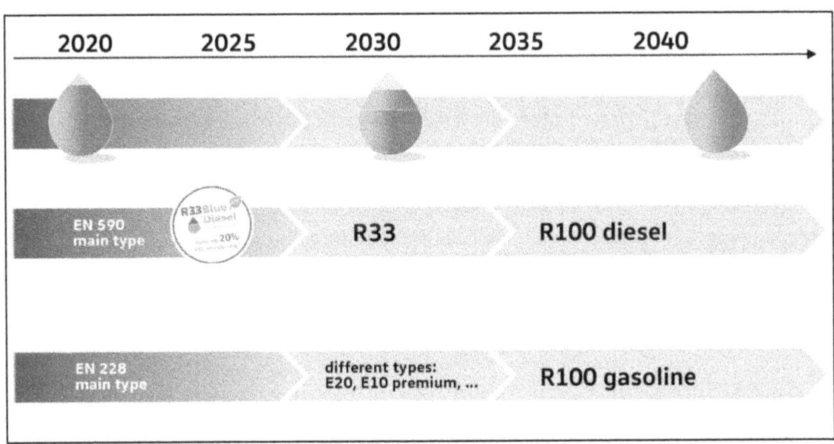

Bild 1: Kraftstoff-Roadmap der Volkswagen AG

Eine Zumischung regenerativer Komponenten innerhalb der Normen EN 590 bzw. EN 228 bis zu 33% gelingt technisch. Ein vielversprechender Kraftstoff für die nächste Dekade ist E20, da Ethanol bei einer Beimischung von bis zu 20 Prozent systemverträglich ist. Die Herstellung ist für Biorohstoffe der so genannten ersten Generation bereits Stand der Technik. Mehrere Hersteller skalieren aktuell die Ethanolherstellung aus zusätzlichen Rohstoffen auf eine industrielle Technologiereife, um die Anteile am Gesamtverbrauch zu erhöhen. Durch den Einsatz von 20 Prozent Ethanol ergeben sich zusätzliche Freiheitsgrade für die Beimischung regenerativer Kohlenwasserstoffe, die oft eine niedrige Oktanzahl aufweisen. Durch die hohe Oktanzahl des Ethanols wird dieser Nachteil im Blend ausgeglichen. E20 kann im Jahr 2030 zu 100% regene-

rativ hergestellt werden. Eine Nutzungserlaubnis kann allerdings erst bei Vorliegen einer verbindlichen Norm ausgesprochen werden. Bild 1 zeigt schematisch die Kraftstoff-Roadmap der Volkswagen AG, welche einen realistischen Hochlauf der Produktionskapazitäten berücksichtigt und die maximale Einsatzfähigkeit im Fahrzeugbestand ermöglicht.
Eine aussichtsreiche Option zur Herstellung der Kohlenwasserstoff-Grundware ist Methanol to Gasoline. Bei diesem Verfahren wird aus Methanol an einem Zeolith-Katalysator ein Kohlenwasserstoffstrom erzeugt, der als Benzinkomponente eingesetzt werden kann. Im C3-Mobility-Projekt („Closed Carbon Cycle") wurde eine hinreichende Menge von MtG-Kraftstoff hergestellt, um umfangreiche Untersuchungen im Labor, an Motoren und Fahrzeugen unterschiedlicher Hersteller vorzunehmen [2] und die Skalierbarkeit der Herstellung auf den zweistelligen Tonnenmaßstab nachzuweisen.

2. Ausgewählte Ergebnisse der Untersuchungen an Motoren und Fahrzeugen der Volkswagen AG im C3-Mobility-Projekt

Im C3-Mobility-Projekt wurden bei Volkswagen verschiedene MTG-Mischungen mit 10 bzw. 20% Ethanol an Motor- und Rollenprüfständen untersucht. Es wurden neben dem Einsatz in unmodifizierten Motoren auch eine Software-Modellapplikation (inklusive der Kraftstofferkennung durch einen Sensor) und eine Motorhardware-Anpassung vorgenommen. Die Emissionen von Schadstoffen und Treibhausgasen bei Einsatz dieser Kraftstoffe wurden im WLTP sowie unter aggressiven Fahrbedingungen bewertet. Die vollständigen Ergebnisse der Untersuchungen an Motoren und Fahrzeugen der Volkswagen AG werden im Abschlussbericht des Projektes [3] zusammengefasst. In der vorliegenden Unterlage werden exemplarisch Emissionen und Verbrauch von MTG – Mischungen mit 10 bzw. 20% Ethanol im WLTP-Zyklus am Fahrzeugprüfstand berichtet.

Untersuchte Kraftstoffqualitäten

Im Projekt wurden nacheinander zwei verschiedenen MtG-Grundqualitäten, Batch 1 (B1) und Batch 2 (B2), hergestellt. Diese wiesen jeweils eine Oktanzahl von ca. 92 auf. Da die getesteten Motoren auf eine minimale Oktanzahl von 95 ausgelegt sind, wurde der Betrieb mit den reinen Grundqualitäten aufgrund der niedrigen Oktanzahl als nicht sinnvoll erachtet und der reine MtG-Kraftstoff nicht getestet. Aus den Grundqualitäten sind durch Zugabe von Ethanol B1 E10 und B2 E10 sowie eine E20-Variante B2 E20 geblendet worden. Diese Kraftstoffe und unterschiedliche Referenzkraftstoffe (VW E10, VW E20 und EN228 Worst Case) wurden in diversen Tests eingesetzt. Der EN228 Worst Case bildet das schlechte 10% Perzentil der Marktkraftstoffe der EN228 ab. Insbesondere die Rußneigung ist bei diesem Kraftstoff besonders hoch. Bild 2 zeigt ausgewählte Eigenschaften der eingesetzten Kraftstoffe.

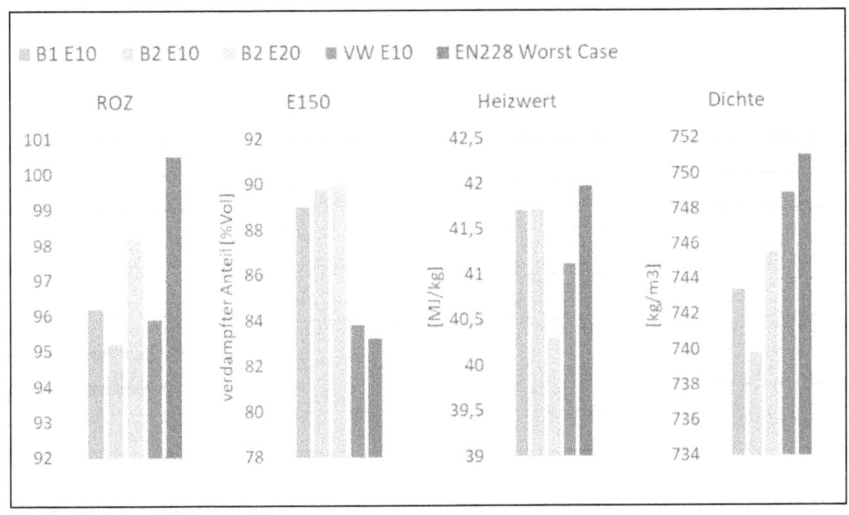

Bild 2: Ausgewählte Eigenschaften der untersuchten Kraftstoffe

Eingesetzte Motoren und Fahrzeuge

Es wurden aktuelle 1,5L-Motoren des Typs EA 211 evo eingesetzt. Für die Untersuchungen am Rollenprüfstand wurden diese in Golf und Passat verbaut. Beim Verbau werden drei Varianten unterschieden:

- Einsatz des unmodifizierten Serienstandes
- Integration eines Kraftstoffsensors, welcher in der Lage ist, E20-Kraftstoff zu erkennen, in Kombination mit einer Umschaltung auf einen E20-Kennfeldsatz in der Motorsteuerung
- Einbau modifizierter Kolben zur Erhöhung der Verdichtung von 10,5 auf 11,6 und entsprechende Anpassung der Motorsteuerung (s. Bild 3)

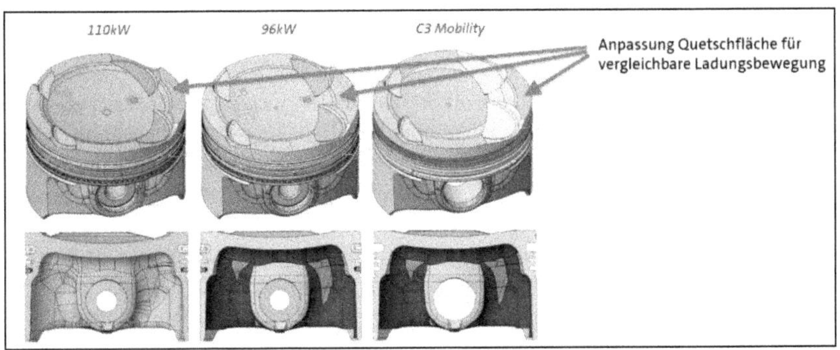

Bild 3: Hardwaremodifikationen für den ausschließlichen Einsatz von E20 mit ROZ > 98

Darüber hinaus sind alle Versuchsträger mit entsprechender Messtechnik und insbesondere Abgasentnahmestellen ausgerüstet. Im weiteren Verlauf der Ausarbeitung werden Rohemissionen adressiert. Diese basieren auf der Abgasentnahmestelle zwischen Abgasturbolader und Abgasnachbehandlung, sowohl am Motor als auch im Fahrzeug. Im Fahrzeug werden insbesondere für CO_2-Emissionen auch Tailpipe-Emissionen gemessen.

Bild 4 stellt die untersuchten Kraftstoffe und Fahrzeugkonzepte schematisch dar.

Technologiestufe	Kraftstoff	Versuchsinhalt	Konzept
1 MtG als Drop-In Kraftstoff	MtG E10	Emissionsbewertung: - Motorprüfstand - Rollenprüfstand	
2 MtG E20 optimierter Motorbetrieb durch Kraftstofferkennung	MtG E10 MtG E20	Implementierung Kraftstoffsensor Emissionsbewertung: - Motorprüfstand - Rollenprüfstand	+ Fuel sensor
3 MtG E20 Hard-und softwareoptimiertes Fahrzeug	MtG E10 MtG E20	HighEpsilon-Motorkonzept Emissionsbewertung: - Motorprüfstand - Rollenprüfstand	+

Bild 4: Übersicht Motor- und Fahrzeugkonzepte

Ausgewählte Emissionsergebnisse

In Abbildung 5 sind (jeweils die ersten beiden Balken) die Ergebnisse für den unmodifizierten Motor aufgetragen. Hier wird als Referenz der Borderline (Worst Case – Kraftstoff) angezogen und mit dem B10 werden die zweiten (besseren) Batches von MtG aus dem Versuch dargestellt. In den folgenden drei Balken wird die Applikation mit der höheren Verdichtung dem gegenüber dargestellt. Als Kraftstoffe werden der Borderline-Kraftstoff, der MtG E10 und der MtG E20 verglichen. Für diese Darstellung wurde der grenzwertige Kraftstoff ausgewählt, da er eine Oktanzahl aufweist, welche die Vorteile der höheren Verdichtung nutzen kann und keine Verschiebungen des Betriebs zu schlechteren Wirkungsgraden erfordert. Dies wäre beim „VW E10“ der Fall.
Bei den Kohlenwasserstoffemissionen zeigen die MtG-Kraftstoffe einen leichten Trend zu geringeren Emissionen. B1 E10 ist im Rahmen der Standardabweichung gleich

dem Referenzkraftstoff. Beide B2 Varianten zeigen leichte Tendenzen von ca. 4% geringeren HC-Emissionen. Die MtG-Kraftstoffe B1 E10 sowie B2 E10 befinden sich bei den Kohlenstoffmonooxidemissionen auf dem gleichen Niveau wie der Referenzkraftstoff. B2 E20 emittiert ca. 6% mehr CO im Fahrzeugtest. B1 E10 sowie B2 E10 emittieren in etwa gleich viel NO_x wie der Referenzkraftstoff. Leichte Vorteile von ca. 6% geringeren NO_x-Emissionen werden bei der MtG E20 Variante gemessen. Bei den PN-Emissionen kann eine deutlich geringere Partikelbildungsneigung der MtG-Kraftstoffe gegenüber dem Referenzkraftstoff festgestellt werden. Außerdem scheint die zweite Grundware im Vergleich zur ersten eine verbesserte PN-Reduktion von ca. 45% normiert auf den EN228 Worst Case Kraftstoff zu bewirken. Weiterhin kann festgestellt werden, dass ein höherer Ethanolanteil zu einer geringeren Partikelanzahl führt. Die maximale Partikelreduktion von ca. 50% kann beim B2 E20 beobachtet werden. Kohlenstoffdioxid wird im Rahmen der Standardabweichung von beiden MtG E10 Varianten in einem ähnlichen Maße wie vom Referenzkraftstoff emittiert. Die E20 Variante zeigt eine Tendenz zu leicht geringeren CO_2-Emissionen von ca. 1,5%.

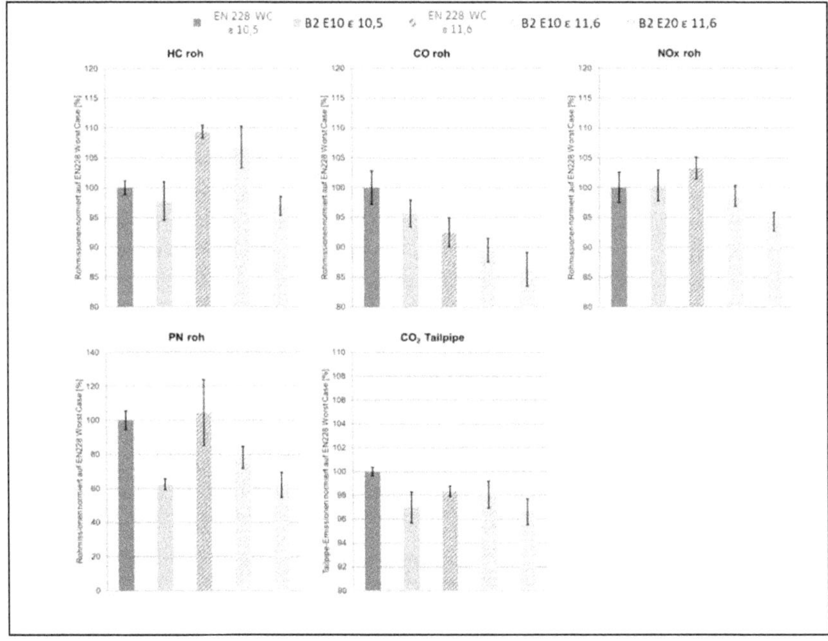

Bild 5: Emissionen und Verbrauch der Fahrzeuge im C3-Mobility-Projekt (WLTP, 23°C)

Der B1 E10 Kraftstoff zeigt im Serienfahrzeug ca. 5% weniger HC-Emissionen, ca. 8% erhöhte CO-Emissionen und unauffällige NOx-Emissionen. Die Partikelemissionen

nehmen um ca. 45% ab. CO_2 ist im Rahmen der Messgenauigkeit ähnlich zum Referenzkraftstoff zu bewerten. B2 E20 zeigt ähnliche Emissionsergebnisse in CO, NOx und PN23. Bei den HC-Emissionen kann eine weitere Verringerung festgestellt werden. Die CO_2-Emissionen des MtG E20 sind signifikant niedriger als B1 E10. Dies kann auf den um 10% erhöhten Ethanolanteil zurückgeführt werden.
Mit dieser CO_2-optimierten Applikation sind CO-, HC- und NOx-Emissionen unauffällig im Vergleich zur Basis-Applikation. Die PN23-Emissionen hingegen steigen leicht an. Die CO_2-Emissionen liegen signifikant um ca. 2% niedriger im Vergleich zum Referenzkraftstoff. Die Einsparungen liegen niedriger als dies erwartet wurde.
Im Rahmen des Projektes wurden auch bei anderen Motor- und Fahrzeugherstellern Untersuchungen an MtG – Kraftstoffen vorgenommen. Darüber hinaus wurde an ausgewählten Kraftstoffsystemen eine hohe (gleich bis besser im Vergleich zu marktüblichem Vergleichskraftstoff) Verträglichkeit für MtG-basierte Kraftstoffe, welche der Kraftstoffnorm EN 228 genügen, nachgewiesen [4].

3. Optionen zur Weiterentwicklung von MtG-Kraftstoffen

Die Minderung des CO_2-Ausstoßes des optimierten Fahrzeugs blieb hinter den Erwartungen zurück. Deshalb wurden bei Volkswagen Untersuchungen mit weiteren Kraftstoffen durchgeführt.
Exemplarisch sind in Tabelle 1 die Ergebnisse dargestellt, die beim Einsatz eines qualitativ verbesserten Kraftstoffes erzielt wurden. Dieser Kraftstoff basiert auf einer fossilen Grundware mit sehr hoher ROZ sowie einem sehr niedrigen Gehalt an schweren Aromaten. Dieser Grundware wurden 20% Ethanol (hohe ROZ, hohe Verdampfungsenthalpie) und ein hochwertiges Additivpaket zugesetzt. Mit diesem Kraftstoff konnten im WLTP Partikelminderungen von über 70% gegenüber der Ausgangsmessung und eine Minderung des CO_2-Ausstoßes um über 5% realisiert werden.
Hanno Krämer [5] hat theoretisch Optionen zur Synthese hoch klopffester Alkan- Komponenten über den Zwischenschritt Isobuten aufgezeigt. Diese Komponenten können problemlos mit MtG geblendet werden, so dass sich eine vergleichbare Qualität ergibt.

Tabelle 1: Emissionen und Verbrauch ausgewählter Fahrzeuge bei Einsatz der C3-Mobility und weiterer Kraftstoffe (WLTP, 23°C)

Technologiestufe	Kraftstoff	HC	CO	NO_x	PN	CO_2
Drop-In E10 (Mittelwert)	B2 E10	1%	2%	0%	-41%	-0,7%
Drop-In E20	B2 E20	-14%	2%	0%	-43%	-1,4%
Angepasste Hardware und Applikation	B2 E20	-3%	-14%	-6%	-39%	-3,5%
Angepasste Hardware und Applikation	E20 gute Qualität	3%	0%	-15%	-77%	-5,3%

4. Zusammenfassung

Das MtG – Verfahren ist geeignet, synthetische Kraftstoffkomponenten zu produzieren, welche konventionellem Basiskraftstoff weitgehend gleichen. Da es möglich ist, diese Komponenten regenerativ herzustellen, kann durch die Nutzung dieses Verfahrens ein treibhausgasneutraler Kraftstoff bereitgestellt werden. Der Einsatz solcher Kraftstoffe ist im Fahrzeugbestand möglich, eine kurzfristige Option zur treibhausgasneutralen Mobilität mit bestehenden Ottofahrzeugen Fahrzeuge liegt somit vor.
Im Rahmen des C3-Mobility-Projekts wurden bei Volkswagen Emissionsuntersuchungen mit MTG-Blends durchgeführt, welche 10 bzw. 20 Prozent Ethanol enthalten. In verschiedenen Emissionstests wurde gezeigt, dass bei Verwendung dieser Blends die Schadstoff- und CO_2 Emissionen in der Größenordnung des Betriebs mit konventionellen EN 228-konformen Kraftstoffen liegen. Es wurden weiterhin gezeigt, dass eine Applikation auf den E20 Kraftstoff u.a. mit einer höheren Verdichtung leichte Vorteile von etwa 3,5% CO_2 ermöglicht. An einem derartig modifizierten Fahrzeug wurde weiterhin ein Kraftstoff mit verbesserter fossiler Grundware vermessen. Sie zeichnet sich durch eine hohe Oktanzahl sowie einen geringen Gehalt an schweren Aromaten aus, welche besonders zur Partikelbildung beitragen. Diese Fahrzeug-Kraftstoff-Kombination zeigt gegenüber dem Ausgangskraftstoff im Standardmotor Partikelminderungen von über 70% sowie eine CO_2-Minderung von über 5%.
Es wird empfohlen, eine schnelle Skalierung des MtG-Verfahrens mit einer gleichzeitigen Verbesserung der Grundqualität zu verbinden.

Literatur

[1] Tatsachen und Zahlen, 84th edition 2020, VDA (only digital edition available)

[2] Kuschel, Gräbner „MtG as an EN 228 Compatible Drop in Fuel“, Pathways to Clean Fuels, virtueller Kongress, FEV 2021

[3] Closed Carbon Cycle Mobility Klimaneutrale Kraftstoffe für den Verkehr der Zukunft, *Abschlussbericht in Erstellung, 2022*

[4] Garbe, Kaszas, Lippmann, Willems, Menger, Rausch, Kolbeck, Döhler, Eitel Kapp, Deeg, Villforth, Mühlstedt, Lucka, Eiden, Heuser, „MtG - Efficiently Reducing fossil CO_2-Emissions of Vehicle Fleet“, Pathways to Clean Fuels, virtueller Kongress, FEV 2021

[5] Krämer, „High octane paraffinic Fuel“, 4. Tagung der FJRG 2021

Methane Concentration Influence on the Ignition Delay of Dual Fuel Combustion Strategy

Rafael Clemente Mallada

Abstract

Dual Fuel Internal Combustion Engines (DFICE) are a promising alternative for substituting heavy duty diesel engines for maritime, mining, agrarian and stationary applications. DFICE ignite by compression a high cetane number fuel to produce a pilot flame that results in the ignition of a high octane number fuel/air-mixture. This combustion strategy aims to obtain the efficiency and torque of Diesel engines producing only a fraction of pollutants and dramatically reducing soot emissions for short injections. In the case of DFICE that use natural gas as substitution fuel it is of vital importance to correctly understand the combustion mechanism in order to assure a complete methane combustion: unburnt methane is estimated to have a global warming potential (GWP) of 28–36 over a 100 years time-lapse.

A first step to research towards achieving a complete combustion is to understand the role of the chemical properties (reactivity) of methane in the ignition process. In fact, although it may result counterintuitive, methane turns out to have a flame inhibitor behaviour and its combustion is modelled in single step chemistry,

$$k_{\mathrm{ov}} = AT^{n}\exp\left(-\frac{E_a}{RT}\right)[\text{ Fuel }]^{a}[\text{ Oxidizer }]^{b}, \qquad (1)$$

by powering its concentration to a small negative exponent ($a \simeq -0.3$) [1].

On the other hand, the ambient methane displace oxygen reducing the oxygen concentration. It is good here to drag our attention to some studies about the influence of $[\,O_2]$ in the ignition delay when using various EGR levels like [2].

The present work varies the concentration of methane in the recharge air of an optically accessible Rapid Compression Machine that is used to recreate the conditions during an engine compression stroke with a volumetric compression ratio of $r = 14$.

Fig. 1: Rapid Compression Machine

Between 10 to 15 mm before the piston reaching the top dead centre, a variable amount of diesel at 1600 bar is injected to act as pilot fuel and ignite the mixture. The hot combustion is detected there by recording at high speed the natural OH* chemiluminescence of the reaction.

Eventually, a vague correlation can be observed that obeys an Arrhenius equation type between ignition delay and the three species' concentrations at the time when the first OH* signal appears.

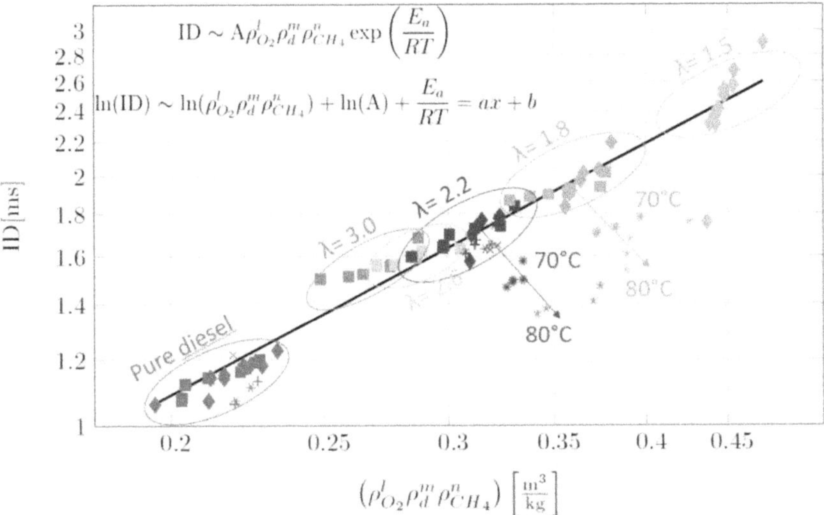

Fig. 2: Ignition delay correlation with species' concentrations for l = -0.9, m= - 0.3 and n = 0.2.

Literature

[1] Charles K., Westbrook and Frederick L. D. (1981). Simplified Reaction Mechanisms for the Oxidation of Hydrocarbon Fuels in Flames. *Combustion Science and Technology*. 27(1-2):31-43 doi:10.1080/00102208108946970

[2] C. A. Idicheriay and L. M. Pickett (2011). Ignition, soot formation, and end-of-combustion transients in diesel combustion under high-EGR conditions. *International Journal of Engine Research*. 12(4):376-392
DOI: 10.1177/1468087411399505

OH* Chemiluminescence and Soot Luminosity Measurement of Two Large-Scale Heavy-Duty Diesel Injectors with Two Identical Cetane Number Fuels, RME and Reference Diesel

Hamidreza Fajri

1. Abstract

In the transition to sustainable mobility, drop-in biofuels are important both in the short- and medium-term transfer process and as a sustainable energy route in the long term. Rapeseed Oil Methyl Ester (RME), which is produced by esterification of the rapeseed oil is analysed along with reference diesel fuel with identical cetane numbers in a constant volume combustion chamber by utilizing two optical methodologies, OH* chemiluminescence and natural flame luminosity (soot). Excluding the influence of cetane number on ignition delay brought the chance to relate fuel volatility and oxygen content to start of the ignition and the lift-off length. Although the results illustrate that RME has a higher liquid penetration and a visible delay regarding the liquid/vapour phase change, this biofuel nevertheless shows a shorter ignition delay when compared to diesel - due to its chemical structure including 10% of oxygen. In the current study, a high-speed camera with an intensifier and a band-pass filter (around 308 nm) is synchronized with a second high-speed camera, which are simultaneously utilized to scrutinize the combustion behaviour of two large-scale heavy-duty marine injectors with an outlet diameter of 300 µm. The two injectors with a conicity factor of 0 and 4 demonstrate geometry effects and show visible diversity in flame initiation and propagation and connect spray cone angle and penetrations to the formation and location of soot and OH* signals. The results shed light upon the fact that more air entrained to the spray due to a wider cone angle of the cylindrical nozzle has a non-negligible influence as it causes faster ignition initiation. The ability to change the ambient properties in the combustion chamber means that ignition delay, the length of the lift-off process, and soot formation can be correlated with the real conditions in the engine cylinder, and influences of individual parameters can be analysed in isolation.

2. Main result

Figure 1 shows an example of diffusion flame of diesel fuel, grey color shows OH* signal and the false color illustrates natural flame (soot) signal. The fuel pressure of the right side image is 100 MPa and the left side is 150 MPa. Ambient gas pressure is 6 MPa and temperature is 923 K for both sides. The times that are written on top sides are regarded as the difference of the current step and the start of injection; however, both indicate a difference of 75 µs from start of ignition. Fuel pressure as presented speeds up the mixture formation process, which accelerates the appearance of OH* and flame signals. As a result, 400 µs is the ignition delay of 150 MPa fuel pressure and 450 µs is a longer ignition delay for 100 MPa fuel pressure. Cyan and magenta colors on figure 1 show flame lift of length and soot lift of length, respectively.

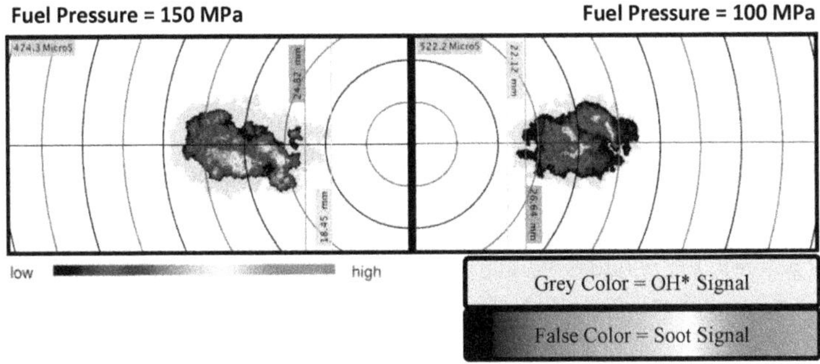

Figure 1: Diffusion flame of diesel fuel in two different pressure levels

2.1. Theory

According to [1, 2], the flame lift-off length could have the following relationship to different parameters:

$$H = CT_a^{-3.74}\rho_a^{-0.85}d^{0.34}U^1Z_{st}^{-1}$$

H (mm) represents lift-off length, C is a tuning constant, T_a (K) is ambient temperature, ρ_a (kg/m^3) ambient density, d (μm) is injector nozzle exit diameter, U (m/s) is injection velocity and Z_{st} is stoichiometric mixture fraction.

Spray propagation, air/fuel mixture formation and therefore exhaust gas emissions are widely influenced by nozzle geometry of diesel injectors [3]. From an experimental point of view, Benajes et al. [4] pointed out that conical nozzles reduce cavitation phenomena and increase discharge coefficient compared to cylindrical nozzles, which finally could influence mixture formation. Apart from the effect of nozzle geometry, alternative fuels have been proven to have advantages in replacing conventional fuel. Within several introduced alternative fuels, RME shows non-toxic, biodegradable and renewable trends with a potential to reduce exhaust emissions [5]. In the sensitivity analysis done by Pastor et al. [6], it was proved that RME and air temperature both delay the first soot formation. On the contrary, injection pressure has a negative influence as it speeds up the first soot formation.

Literature

[1] Siebers, D. and B. Higgins, *Flame lift-off on direct-injection diesel sprays under quiescent conditions.* SAE Transactions, 2001: p. 400-421.

[2] Siebers, D., B. Higgins, and L. Pickett, *Flame lift-off on direct-injection diesel fuel jets: oxygen concentration effects.* SAE Transactions, 2002: p. 1490-1509.

[3] Blessing, M., et al., *Analysis of flow and cavitation phenomena in diesel injection nozzles and its effects on spray and mixture formation.* SAE transactions, 2003: p. 1694-1706.

[4] Benajes, J., et al., *Analysis of the influence of diesel nozzle geometry in the injection rate characteristic.* J. Fluids Eng., 2004. **126**(1): p. 63-71.

[5] Hopp, M., *Untersuchung des Einspritzverhaltens und des thermischen Motorprozesses bei Verwendung von Rapsöl und Rapsmethylester in einem Common-Rail-Dieselmotor.* 2005.

[6] Pastor, J., et al., *Experimental study on RME blends: liquid-phase fuel penetration, chemiluminescence, and soot luminosity in diesel-like conditions.* Energy & Fuels, 2009. **23**(12): p. 5899-5915.

Abschätzung des Flüssiganteils in Sprays mittels DBI unter Berücksichtigung verschiedener Streuungseffekte

Bastian Lehnert

1. Einfluss von Streuungseffekten auf die optische Dichte

Die Kraftstoffeinbringung und Gemischbildung sind essentielle Prozesse in Verbrennungsmotoren. Ein genaues Verständnis der Sprayausbreitung ist daher, unabhängig von der Kraftstoffart, notwendig, um hocheffiziente und saubere Verbrennungsmotoren zu bauen.

Aus diesem Grund wurden im Feld der Sprayuntersuchungen verschiedene numerische und experimentelle Untersuchungen gemacht. Optische Messmethoden sind hierbei ein weit verbreitetes Werkzeug [1]. Insbesondere bei der Erfassung makroskopischer Eigenschaften haben sich verschiedene Messverfahren etabliert, um Eindringtiefe, Winkel und Flächen der Flüssig- und Gasphase zu messen [2-5]. Diese Techniken sind jedoch nicht zur Erfassung quantitativer Daten geeignet.

Auf Grundlage von Ghandi und Heim [6,7] und weiterentwickelt von Westley et al. [6] wurde daher die Messmethode „Diffuse back-illumination imaging" (DBI) entwickelt. Diese Messmethode ermöglicht die Quantifizierung des projizierten Flüssigvolumenanteils (PLVF) innerhalb des Sprays durch Umrechnung der optischen Dichte τ:

$$\frac{I}{I_0} = e^{-\tau} \tag{1}$$

$$\tau * \frac{\pi d^3/6}{C^*_{ext}} = PLVF \tag{2}$$

Mittels des ursprünglich von der Lichtquelle erzeugten Signals I_0 und dem durch die Flüssigphase abgeschwächten Signals I wird in (1) die optische Dichte τ berechnet. Anschließend kann mit dem Tropfendurchmesser d und dem Extinktionskoeffizienten C^*_{ext} der projizierte Flüssigvolumenanteil berechnet werden (2). Durch die Verwendung einer homogenen diffusen Lichtquelle wird zudem Strahlbeugung und somit die Detektion der Gasphase und Temperaturgradienten unterdrückt.

Eine weitere Fehlerquelle, neben der Beugung, bei DBI sind Streuungseffekte. Unter der Annahme, dass Mehrfachstreuungseffekte nur einen geringen Einfluss auf die gemessene optische Dichte haben, wurde dieser Effekt bisher ignoriert. Diese Annahme wurde jedoch nur für eine kollimierte Lichtquelle bewiesen [9]. Da ein wichtiger Aspekt der DBI jedoch die diffuse Lichtquelle ist, gilt diese Annahme nicht.

Ziel dieser Arbeit ist die Erfassung des durch Streuungseffekten hervorgerufenen Fehlers, sowie dessen Korrektur. Es wird hierfür mit einer Monte Carlo-Simulation dieser Fehler simuliert. Bild 1 zeigt Ergebnisse dieser Simulation mit zwei Korrekturkurven für eine kollimierte Lichtquelle und eine Lichtquelle mit 20° Abstrahlwinkel. Es wird über die gemessene optische Dichte (OD mit Streuungseffekten) die reale Optische Dichte (OD ohne Streuungseffekte) aufgetragen.

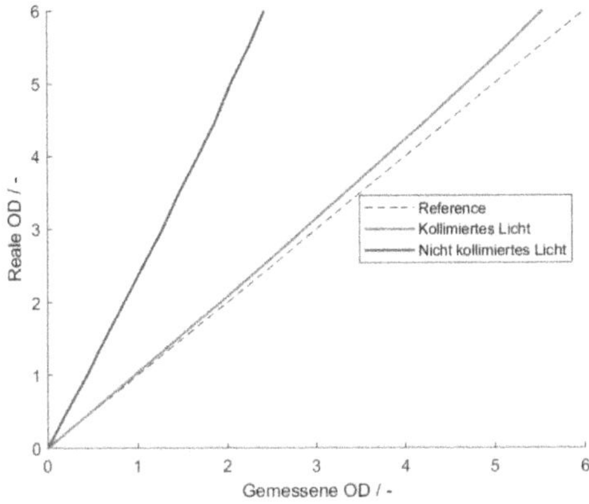

Bild 1: Vergleich von gemessener und realer optischer Dichte für zwei unterschiedliche Lichtquellen

Die Annahme der Streuungsunabhängigkeit bis zu OD 2 kann in Bild 1 für kollimiertes Licht gesehen werden (rot). Für eine nicht kollimierte Lichtquelle wird die optische Dichte jedoch von Anfang an stark unterschätzt. Die Korrekturkurve (blau) zeigt hier eine Steigung > 2. Die Informationen aus Bild 1 werden zur Berechnung von korrigierten ODs verwendet, welche anschließend in die PLVF umgerechnet werden können (Gleichung 2).

1.1 Simulationsaufbau und experimenteller Aufbau

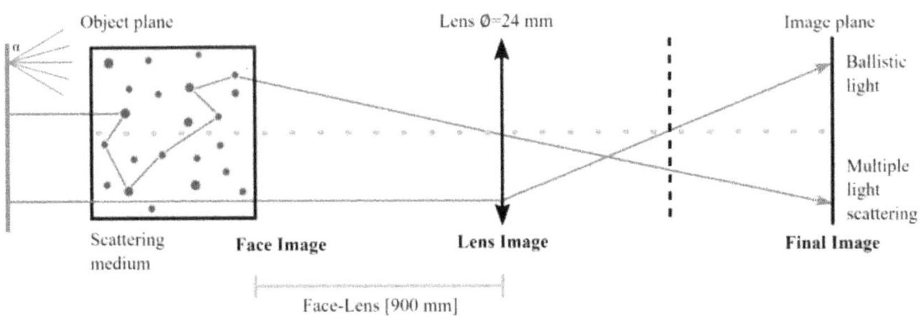

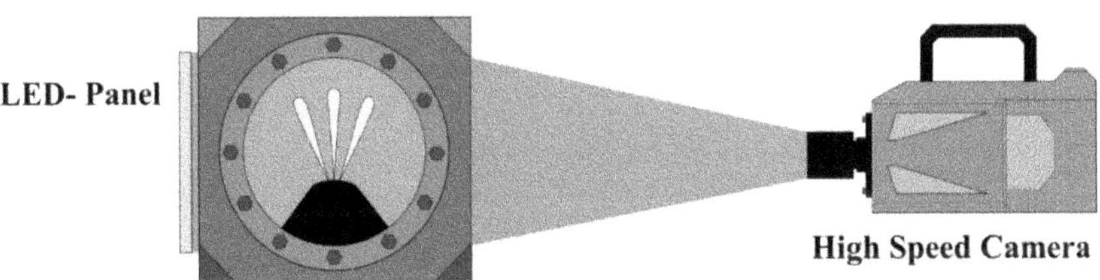

Literatur

[1] Fansler, T. D., Parrish, S. E., Spray measurement technology: a review. Meas. Sci. Technol. 2015, 26, 12002.

[2] Lehnert, B., Conrad, C., Wensing, M., GDI Sprays with up to 200 MPa Fuel Pressure and Comparison of Diesel-like and Gasoline-Like Injector Designs, SAE Technical Paper Series. SAE Powertrains, Fuels & Lubricants Meeting, SEP. 22, 2020. SAE Technical Paper Series, SAE International400 Commonwealth Drive, Warrendale, PA, United States, 2020.

[3] Lazzaro, M., High-Speed Imaging of a Vaporizing GDI Spray: A Comparison between Schlieren, Shadowgraph, DBI and Scattering, SAE Technical Paper Series. WCX SAE World Congress Experience, APR. 21, 2020. SAE Technical Paper Series, SAE International400 Commonwealth Drive, Warrendale, PA, United States, 2020

[4] Riess, S., Weiss, L., Rezaei, J., Peter, A. et al., The liquid penetration of diesel substitutes, in: Payri, R., Margot, X. (Eds.), 28th European Conference Liquid Atomization and Spray Systems: 6th-8th September 2017, Valencia, Spain: ILASS 2017. ILASS2017 - 28th European Conference on Liquid Atomization and Spray Systems, 06.09.2017 - 08.09.2017, Editorial Universidad Politécnica de Valencia, Valencia 2017.

[5] Manin, J., Jung, Y., Skeen, S. A., Pickett, L. M. et al., Experimental Characterization of DI Gasoline Injection Processes, SAE Technical Paper Series. JSAE/SAE 2015 International Powertrains, Fuels & Lubricants Meeting, SEP. 01, 2015. SAE Technical Paper Series, SAE International400 Commonwealth Drive, Warrendale, PA, United States, 2015.

[6] Ghandhi, J. B., Heim, D. M., An optimized optical system for backlit imaging. The Review of scientific instruments 2009, 80, 56105.

[7] Westlye, F. R., Penney, K., Ivarsson, A., Pickett, L. M. et al., Diffuse back-illumination setup for high temporally resolved extinction imaging. Applied optics 2017, 56, 5028–5038.

[8] Berrocal, E., Sedarsky, D. L., Paciaroni, M. E., Meglinski, I. V. et al., Laser light scattering in turbid media Part I: Experimental and simulated results for the spatial intensity distribution. Opt. Express, OE 2007, 15, 10649–10665.

Reaction mechanisms development for the sustainable fuel combustion

Solmaz Nadiri

1. Motivation

The growing concern over global warming has led to an increase in the development of carbon-neutral technologies. It is becoming increasingly important to find alternative fuels that have low to zero carbon dioxide (CO_2) emissions.
The combustion of sustainable fuels involves mixtures of fairly large molecules and relatively unselective chemistry, leading to complex product mixtures. These corresponding reaction networks are quite complex since each molecule in the feed can form many isometric intermediates and a variety of byproducts in addition to its major product. Building accurate reaction mechanisms for these complicated systems is challenging, and the methodology is still under development.
In this work, the combustion properties of two promising alternative fuels, namely alcohol and liquified natural gas (LNG) have been investigated to develop a comprehensive reaction mechanism for each fuel. Consequently, sensitivity analysis has been used to determine the most sensitive reactions influencing the ignition of each fuel under the studied temperatures and pressures.

2. Numerical simulation

An in-house code of Cantera [1] version 2.4.0 was used for all simulations performed in this study. To represent the mixtures containing multiple species and compute the thermodynamic and kinetic properties of the mixtures, class Solution was used.
To reduce the computational time of the numerical simulations, further investigations were conducted on the carbon-nitrogen cross reactions generated by RMG based on the sensitivity analysis and rate of production analysis.
It was found that the rate constants of the elementary reactions from the literature were not up-to-date or numerically determined, which leads to higher uncertainty in the studied temperature and pressure ranges. Therefore, literature mechanisms containing those most sensitive reactions were selected to compare the rate constants applied to identify the most appropriate one for predicting experimental IDTs. Applying the trial-and-error method, where the rate constant of the aforementioned sensitive reactions was substituted one by one, the performance of the mechanism has been comprehensively evaluated.

3. Results

The newly developed mechanism for LNG was validated first against the measured IDTs categorized according to the methane number (MN) [2], cf. Fig. 1. Moreover, the developed mechanism for propanol and butanol was validated first against the IDTs measured in the HPST (this work) and RCM (previous works in our group) [3]. Fig. 2 depicts the comparison between simulations and experimental results measured in

RCM at ϕ = 0.25, 0.5 and 0.9 for propanol and butanol isomers, respectively. The developed mechanism predicts the experimental trends precisely for each fuel at all the equivalence ratios, even at a very lean condition with a very small amount of fuels.

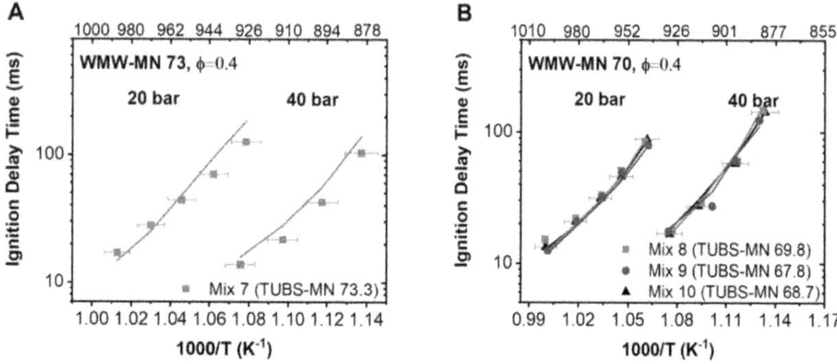

Figure 1: IDTs of LNG mixtures measured in the RCM at 20 and 40 bar, ϕ = 0.4. (Symbols: experimental points; Lines: model predictions)

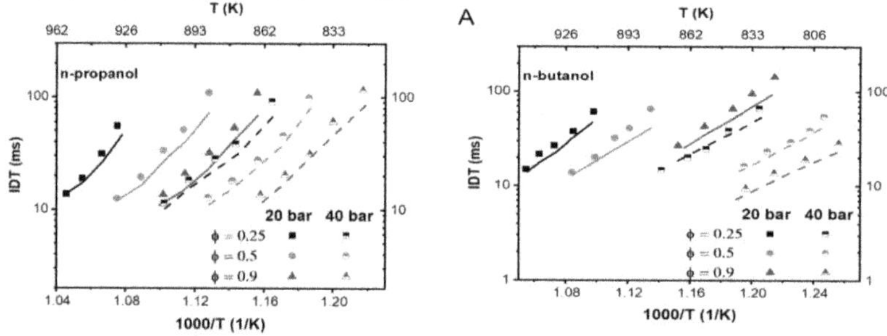

Figure 2: IDTs of butanol and propanol measured in the RCM at 20 and 40 bar and ϕ = 0.25, 0.5 and 0.9. (Symbols: experimental points; Lines: model predictions)

References

[1] Goodwin DG, Moffat HK, Speth RL. Cantera: An object-oriented software toolkit for chemical kinetics, thermodynamics, and transport processes. Pasadena, CA: Caltech; 2009.

[2] Nadiri, S., Agarwal, S., He, X., Kühne, U., Fernandes, R., & Shu, B. (2021). Development of the chemical kinetic mechanism and modeling study on the ignition delay of liquefied natural gas (LNG) at intermediate to high temperatures and high pressures. Fuel, 302, 121137.

[3] Nadiri, S., Zimmermann, P., Sane, L., Fernandes, R., Dinkelacker, F., & Shu, B. (2021). Kinetic Modeling Study on the Combustion Characterization of Synthetic C3 and C4 Alcohols for Lean Premixed Prevaporized Combustion. Energies, 14(17), 5473.

Mixture Formation and Ignition of Dodecane and $OME_{3\text{-}5}$

Lukas Strauß, Sebastian Rieß, Michael Wensing

1. Introduction

As the CI-engine can still compete with new technologies when operated with sustainable e-fuels, spray characterization is key to reach optimum efficiency and close to zero emission. In this work, $OME_{3\text{-}5}$-mix (OME) is compared with dodecane (DOD) as diesel-like reference fuel in terms of mixture formation and ignition.

2. Experimental setup / Methodology

Since the combustion chamber of a diesel engine is very engine-specific and difficult to access visually, the experiments for the spray investigation are carried out at an optically accessible high temperature and high pressure constant volume injection chamber. The test bench is continuously scavenged with gas mixtures, which can be freely adjusted in terms of gas composition (N_2, O_2), temperature, and pressure. A research fuel system provides the required fuel pressure. The experiments are carried out with an ECN Spray A-3 single hole injector. According to ECN definition, following operating conditions are set:

$\rho_{ambient}$ [kg/m³]	$T_{ambient}$ [K]	p_{fuel} [bar]	T_{fuel} [K]	O_2-content [vol.-%]
22,8	900	1500	363	0 / 21

To obtain data about the fuel spray and its mixture, high speed imaging techniques are applied. For measuring the liquid phase, the fuel spray is illuminated with LED panels and the Mie scattered light is detected. Further, a typical Schlieren setup with a 532 nm LED is applied to the vessel in order to detect the density gradients and thus the gaseous penetration. For the OH*-measurements a bandpass filter with 315 nm and a high-speed intensifier IRO X of LaVision are used. For each operation point 32 injections are recorded with a Photron SA-Z and evaluated with a self-developed MATLAB-based program. In addition, mass flow rate measurements are performed.

3. Results

For gaseous penetration under non-reacting conditions, it has to be noted that the penetration depth is the same for all fuels because of same ambient conditions, especially same ambient density. As mentioned in relevant literature, the gaseous penetration depth depends not on fuel properties [1,2]. Also, the gaseous cone angle does not change with fuel, it mainly depends on nozzle hole geometry [3,4] and is in the present study nearly constant at a value of 19,5°. Relevant for the liquid penetration depth respectively the evaporation of fuel is the enthalpy introduced via air entrainment. Dodecane is completely evaporated after 10,8 mm axial distance from the nozzle tip and

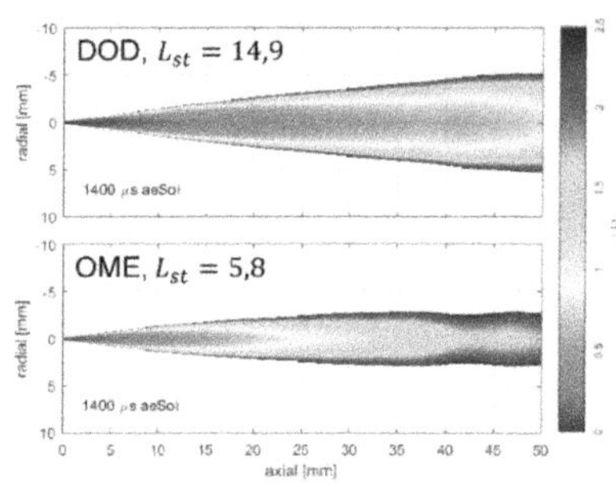

Fig. 1: Air-fuel equivalence ratio λ at 1400 µs after electrical start of injection

OME3-5 after 12,1 mm. The measured data are used to build a reduced dimensional model for penetration depth and fuel-air-ratio as described by Musculus & Kattke, which is based on momentum balance [5]. The measured angle was adjusted slightly so that the measured penetration depths match the modelled penetration depths. The developed model can then be used to calculate the mass ratio of fuel to ambient gas for each location after start of injection. One can see that the fuel to ambient air mass ratio is nearly the same for both fuels, which can be explained with the momentum balance. However, including the stoichiometric air requirement shows that OME3-5 forms a much leaner mixture than dodecane (see Fig. 1). The main reason for this is the chemical structure of the fuels. OME3-5 has impressive 47 wt.% molecularly bound oxygen. It has to be mentioned, that the form of the OME3-5 spray does not change. In the boundary areas, it gets so lean, that it is no longer displayed in Fig. 1. The measurements of the OH*-chemiluminescence show that the ignition events, despite the different mixture quality, are very similar in terms of time, which is also confirmed by the very similar cetane number. However, the lift-off length of the OME3-5 flame is by a lot longer than for the dodecane combustion (Fig. 2), which is quite noticeable as we know from other measurements, that there usually is a linear correlation between the ignition delay and the lift-off length.

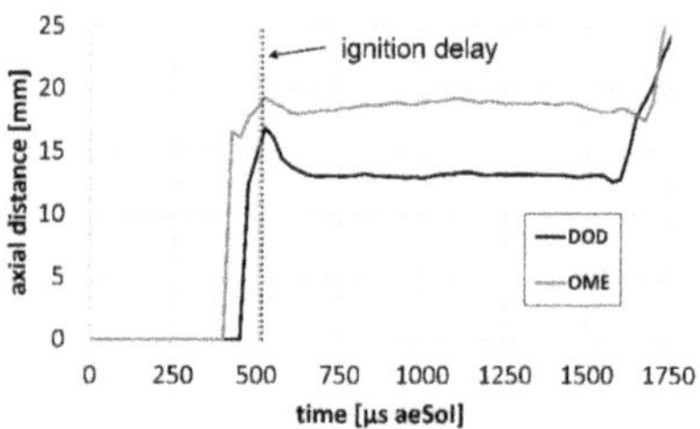

Fig. 2: Ignition delay and OH lift-off length*

4. Conclusions & Outlook

Taking all measurements and models into account, it can be concluded, that OME3-5 behaves like dodecane in terms of mass distribution in the spray, but the air-fuel equivalence ratio is much leaner. In terms of combustion, it can be noted that the temporal ignition delay is very similar, but the local one is not. In order to understand the processes of energy conversion more precisely, the next step will be to have a closer look at the combustion. It will be investigated in which temporal sequence which species appear and disappear to find out, why OME3-5 burns fundamentally differently.

Literature

[1] Desantes, J., Payri, R., Salvador, F., and Soare, V., SAE Technical Paper 2005-01-0913, 2005.

[2] Naber, J. and Siebers, D., SAE Technical Paper 960034, 1996.

[3] Hiroyasu, H. and Arai, M., SAE Technical Paper 900475, 1990.

[4] Siebers, D., SAE Technical Paper 1999-01-0528, 1999.

[5] Musculus, M. and Kattke, K., SAE Int. J. Engines 2(1):1170-1193, 2009.

Autorenverzeichnis

Dr.-Ing. Enrico Bärow, Woodward L'Orange GmbH, Stuttgart

Dipl.-Ing. Ingmar Berger, Woodward L'Orange GmbH, Stuttgart

Dr. rer. pol. David Bothe, Frontier Economics Ltd., Köln

Dr.-Ing. Stefan Buhl, MAN Truck & Bus SE, Nürnberg

Prof. Dr. med. Jürgen Bünger, Institut für Prävention und Arbeitsmedizin der Deutschen Gesetzlichen Unfallversicherung, Institut der Ruhr-Universität Bochum (IPA), Bochum

Patrick Burkardt, M.Sc., Lehrstuhl für Thermodynamik mobiler Energiewandlungssysteme (TME) der RWTH Aachen University

Gabriel Costa de Paiva, DBFZ Deutsches Biomasseforschungszentrum gemeinnützige GmbH, Leipzig

Prof. Dr. Friedrich Dinkelacker, Institut für Technische Verbrennung, Leibniz Universität Hannover

Dipl.-Ing. Frank Dünnebeil, ifeu – Institut für Energie- und Umweltforschung Heidelberg GmbH

Simon Eiden, M.Sc., Tec4Fuels GmbH, Herzogenrath

Hamidreza Fajri, M.Sc., Professur für Technische Thermodynamik, Friedrich-Alexander-Universität Erlangen Nürnberg

Maximilian Fleischmann, M.Sc., Lehrstuhl für Thermodynamik mobiler Energiewandlungssysteme (TME) der RWTH Aachen University

Prof. Dr. Thomas Garbe, Volkswagen AG, Wolfsburg und Hochschule Coburg

Dipl.-Ing. Christian Honecker, Lehrstuhl für Thermodynamik mobiler Energiewandlungssysteme (TME) der RWTH Aachen University

Dipl.-Ing. William Kaszas, Volkswagen Group Services, Wolfsburg

Prof. Dr. Jürgen Krahl, Präsident der Technischen Hochschule Ostwestfalen-Lippe, Lemgo

Alexander Knafl, Ph.D., MAN Energy Solutions SE, Augsburg

Dr.-Ing. Hanno Krämer, Audi AG, Ingolstadt

Dr.-Ing. Ulrich Kramer, FVV e.V., Frankfurt und Ford-Werke GmbH, Köln

Chandra Kanth Kosuru, M.Sc., Tec4Fuels GmbH, Herzogenrath

Bastian Lehnert, M.Sc., Professur für Technische Thermodynamik, Friedrich-Alexander-Universität Erlangen Nürnberg

Dipl.-Ing. Bastian Lehrheuer, Lehrstuhl für Thermodynamik mobiler Energiewandlungssysteme (TME) der RWTH Aachen University

Prof. Dr. phil. Josef Löffl, Institut für Wissenschaftsdialog, Technische Hochschule Ostwestfalen-Lippe, Lemgo

Dr.-Ing. Klaus Lucka, Tec4Fuels GmbH, Herzogenrath

Rafael Clemente Mallada, M.Sc., Professur für Technische Thermodynamik, Friedrich-Alexander-Universität Erlangen Nürnberg

Dr.-Ing. Franziska Müller-Langer, DBFZ Deutsches Biomasseforschungszentrum gemeinnützige GmbH, Leipzig

Prof. Dr.-Ing. Axel Munack, ehemals: Thünen-Institut für Agrartechnologie, Braunschweig

Solmaz Nadiri, M.Sc., AG Reaktionskinetik, Dept. Physikalische Chemie, Physikalisch-Technische Bundesanstalt (PTB), Braunschweig

Karin Naumann, DBFZ Deutsches Biomasseforschungszentrum gemeinnützige GmbH, Leipzig

Prof. Dr.-Ing. Stefan Pischinger, Lehrstuhl für Thermodynamik mobiler Energiewandlungssysteme (TME) der RWTH Aachen University

Petra Rektorik, M.Sc., MAN Energy Solutions SE, Augsburg

Dr.-Ing. Sebastian Rieß, Professur für Technische Thermodynamik, Friedrich-Alexander-Universität Erlangen Nürnberg

Prof. Dr. Wolfgang Ruck, Leuphana Universität Lüneburg

Jörg Schröder, DBFZ Deutsches Biomasseforschungszentrum gemeinnützige GmbH, Leipzig

Dipl.-Ing. Martin Schüttenhelm, Volkswagen AG, Wolfsburg

Dipl.-Ing. Markus Send, Audi AG, Ingolstadt

Lukas Strauß, M.Sc., Professur für Technische Thermodynamik, Friedrich-Alexander-Universität Erlangen Nürnberg

Julian Türck, M.Sc., Tecosol GmbH, Ochsenfurt

Dr. Lukas Weiß, Friedrich-Alexander-Universität Erlangen-Nürnberg

Prof. Dr.-Ing. Michael Wensing, Professur für Technische Thermodynamik, Friedrich-Alexander-Universität Erlangen Nürnberg

Dr.-Ing. Karsten Wilbrand, Shell Global Solutions (Deutschland) GmbH, Hamburg

Dr.-Ing. Michael Willmann, Woodward L'Orange GmbH, Stuttgart

Dr.-Ing. Johann Wloka, MAN Energy Solutions SE, Augsburg

Birgit Maria Wöber, CNG-Club e.V., München

Mitglieder der Fuels Joint Research Group (FJRG)

Prof. Dr. med. Jürgen Bünger, IPA Bochum

Prof. Dr. Friedrich Dinkelacker, Universität Hannover

Prof. Dr.-Ing. Peter Eilts, TU Braunschweig

Prof. Dr. Ravi Fernandes, PTB Braunschweig

Prof. Dr.-Ing. Karl Huber, TH Ingolstadt

Prof. Dr. Jürgen Krahl, TH OWL, Lemgo

Dr. Klaus Lucka, TEC4FUELS GmbH Aachen

Prof. Dr.-Ing. Axel Munack, Rötgesbüttel

Prof. Dr.-Ing. Thomas Schulte, iFE der TH OWL, Lemgo

Prof. Dr.-Ing. Dr. h c. Helmut Tschöke, Universität Magdeburg

Prof. Dr.-Ing. Michael Wensing, FAU Erlangen-Nürnberg

Prof. Dr.-Ing. Gennadi Zikoridse, HS Technik und Wirtschaft Dresden

Angaben zu den Arbeitsgebieten der Mitglieder und zu Publikationen der FJRG sind zu finden unter: www.fuels-jrg.de

www.ingramcontent.com/pod-product-compliance
Ingram Content Group UK Ltd.
Pitfield, Milton Keynes, MK11 3LW, UK
UKHW021653190726
13853UKWH00001B/231

9 783736 976245